Common Core Subject Test Mathematics Grade 5

Student Practice Workbook

+ Two Full-Length Common Core Math Tests

Math Notion

www.MathNotion.com

Common Core Subject Test Mathematics Grade 5

Common Core Subject Test Mathematics Grade 5

Published in the United State of America By

The Math Notion

Web: WWW.MathNotion.com

Email: info@Mathnotion.com

Copyright © 2021 by the Math Notion. All rights reserved. No part of this publication may be reproduced, stored in a retrieval system, or transmitted in any form or by any means, electronic, mechanical, photocopying, recording, scanning, or otherwise, except as permitted under Section 107 or 108 of the 1976 United States Copyright Ac, without permission of the author.

All inquiries should be addressed to the Math Notion.

ISBN: 978-1-63620-079-8

Common Core Subject Test Mathematics Grade 5

The Math Notion

Michael Smith has been a math instructor for over a decade now. He launched the Math Notion. Since 2006, we have devoted our time to both teaching and developing exceptional math learning materials. As a test prep company, we have worked with thousands of students. We have used the feedback of our students to develop a unique study program that can be used by students to drastically improve their math scores fast and effectively. We have more than a thousand Math learning books including:

– **SAT Math Prep**

– **ACT Math Prep**

– **SSAT/ISEE Math Prep**

–**Mathematics Prep Grade 3 to 8**

– **Common Core Math Prep**

–**many Math Education Workbooks, Study Guides, Practice and Exercise Books**

As an experienced Math test preparation company, we have helped many students raise their standardized test scores—and attend the colleges of their dreams: We tutor online and in person, we teach students in large groups, and we provide training materials and textbooks through our website and through Amazon.

You can contact us via email at:

info@Mathnotion.com

Common Core Subject Test Mathematics Grade 5

Get the Targeted Practice You Need to Ace the Common Core Math Test!

Common Core Subject Test Mathematics Grade 5 includes easy-to-follow instructions, helpful examples, and plenty of math practice problems to assist students to master each concept, brush up their problem-solving skills, and create confidence.

The Common Core math practice book provides numerous opportunities to evaluate basic skills along with abundant remediation and intervention activities. It is a skill that permits you to quickly master intricate information and produce better leads in less time.

Students can boost their test-taking skills by taking the book's two practice Common Core Math exams. All test questions answered and explained in detail.

Important Features of the 5th grade Common Core Math Book:

- A **complete review** of Common Core math test topics,
- Over 2,500 practice problems covering all topics tested,
- The most important concepts you need to know,
- Clear and concise, easy-to-follow sections,
- Well designed for enhanced learning and interest,
- Hands-on experience with all question types,
- **2 full-length practice tests** with detailed answer explanations,
- Cost-Effective Pricing,

Powerful math exercises to help you avoid traps and pacing yourself to beat the Common Core test. Students will gain valuable experience and raise their confidence by taking 5th grade math practice tests, learning about test structure, and gaining a deeper understanding of what is tested on the Common Core math grade 5. If ever there was a book to respond to the pressure to increase students' test scores, this is it.

Common Core Subject Test Mathematics Grade 5

WWW.MathNotion.COM

… So Much More Online!

- ✓ FREE Math Lessons
- ✓ More Math Learning Books!
- ✓ Mathematics Worksheets
- ✓ Online Math Tutors

For a PDF Version of This Book

Please Visit WWW.MathNotion.com

Common Core Subject Test Mathematics Grade 5

Contents

Chapter 1 : Place Values and Number Sense 11
   Place Values 12
   Comparing and Ordering Numbers 13
   Numbers in Word Form 14
   Roman Numerals 15
   Rounding Numbers 16
   Odd or Even 17
   Repeating Patterns 18
   Growing Patterns 19
   Patterns: Numbers 20
   Answers of Worksheets 21

Chapter 2 : Whole Number Operations 23
   Adding Whole Numbers 24
   Subtracting Whole Numbers 25
   Multiplying Whole Numbers 26
   Dividing Hundreds 27
   Long Division by Two Digits 28
   Division with Remainders 28
   Rounding Whole Numbers 29
   Whole Number Estimation 30
   Answers of Worksheets 31

Chapter 3 : Number Theory 33
   Factoring Numbers 34
   Prime Factorization 34
   Divisibility Rules 35
   Greatest Common Factor 36
   Least Common Multiple 36
   Answers of Worksheets 37

Chapter 4 : Fractions and Mixed Numbers 39
   Simplifying Fractions 40
   Like Denominators 41
   Compare Fractions with Like Denominators 43
   More Than Two Fractions with Like Denominators 44
   Unlike Denominators 45
   Ordering Fractions 47

Common Core Subject Test Mathematics Grade 5

Denominators of 10, 100, and 1000 48
Fractions to Mixed Numbers 50
Mixed Numbers to Fractions 51
Add and Subtract Mixed Numbers 52
Answers of Worksheets 53

Chapter 5 : Decimals **57**
Adding and Subtracting Decimals 58
Multiplying and Dividing Decimals 59
Rounding Decimals 60
Comparing Decimals 61
Answers of Worksheets 62

Chapter 6 : Ratios and Rates **63**
Simplifying Ratios 64
Writing Ratios 64
Create a Proportion 65
Proportional Ratios 65
Similar Figures 66
Word Problems 67
Answers of Worksheets 69

Chapter 7 : Measurement **71**
Reference Measurement Units 72
Metric Length Units 73
Customary Length Units 73
Metric Capacity Units 74
Customary Capacity Units 74
Metric Weight and Mass Units 75
Customary Weight and Mass Units 75
Temperature Units 76
Time 77
Money Amounts 78
Money: Word Problems 79
Answers of Worksheets 80

Chapter 8 : Algebraic Thinking **82**
Finding Rules 83
Algebraic Word Problems 84
Evaluate Expressions 85
Variables and Expressions 86
Answers of Worksheets 87

WWW.MathNotion.Com

Common Core Subject Test Mathematics Grade 5

Chapter 9 : Geometric 89
- Identifying Angles 90
- Estimate Angle Measurements... 91
- Measure Angles with a Protractor 92
- Polygon Names 93
- Classify Triangles 94
- Parallel Sides in Quadrilaterals... 95
- Identify Rectangles 96
- Perimeter: Find the Missing Side Lengths.. 97
- Perimeter and Area of Squares... 98
- Perimeter and Area of rectangles 99
- Find the Area or Missing Side Length of a Rectangle............................100
- Area and Perimeter: Word Problems...101
- Circumference, Diameter, and Radius 102
- Volume of Cubes and Rectangle Prisms 103
- Answers of Worksheets 104

Chapter 10 : Three-Dimensional Figures 106
- Identify Three–Dimensional Figures...107
- Count Vertices, Edges, and Faces 108
- Identify Faces of Three–Dimensional Figures 109
- Answers of Worksheets 110

Chapter 11 : Symmetry and Transformations 111
- Line Segments 112
- Identify Lines of Symmetry 113
- Count Lines of Symmetry 114
- Parallel, Perpendicular and Intersecting Lines 115
- Answers of Worksheets 116

Chapter 12 : Data Graphs, and Statistics..117
- Mean and Median 118
- Mode and Range 119
- Graph Points on a Coordinate Plane 120
- Bar Graph 121
- Tally and Pictographs 122
- Dot plots 123
- Line Graphs..124
- Stem–And–Leaf Plot 125
- Scatter Plots 126
- Probability Problems 127

Common Core Subject Test Mathematics Grade 5

Answers of Worksheets 128
Chapter 13 : Common Core Math Practice Tests **131**
Common Core GRADE 5 MAHEMATICS REFRENCE MATERIALS 133
Common Core Practice Test 1 135
Common Core Practice Test 2 149
Chapter 14 : Answers and Explanations **163**
Answer Key 163
Practice Test 1 165
Practice Test 2 171

Common Core Subject Test Mathematics Grade 5

Chapter 1 : Place Values and Number Sense

Topics that you'll learn in this chapter:

- ✓ Place Values,
- ✓ Compare and Ordering Numbers,
- ✓ Numbers in Word Form,
- ✓ Roman Numerals,
- ✓ Rounding Numbers,
- ✓ Odd or Even,
- ✓ Repeating Patterns,
- ✓ Growing Patterns,
- ✓ Patterns: Numbers,

Common Core Subject Test Mathematics Grade 5

Place Values

✎ Write numbers in expanded form.

1) Sixty–two   ___ + ___

2) fifty–six   ___ + ___

3) thirty–one   ___ + ___

4) forty–five   ___ + ___

5) twenty–eight   ___ + ___

✎ Circle the correct choice.

6) The 6 in 56 is in the

        Ones place   tens place   hundreds place

7) The 2 in 25 is in the

        Ones place   tens place   hundreds place

8) The 9 in 918 is in the

        Ones place   tens place   hundreds place

9) The 3 in 537 is in the

        Ones place   tens place   hundreds place

10) The 9 in 289 is in the

        Ones place   tens place   hundreds place

WWW.MathNotion.Com

Common Core Subject Test Mathematics Grade 5

Comparing and Ordering Numbers

✎ Use less than, equal to or greater than.

1) 31 _____ 33
2) 57 _____ 49
3) 92 _____ 88
4) 76 _____ 67
5) 43 _____ 43
6) 54 _____ 46
7) 97 _____ 88

8) 42 _____ 36
9) 55 _____ 55
10) 57 _____ 75
11) 28 _____ 38
12) 19 _____ 15
13) 82 _____ 90
14) 78 _____ 84

✎ Order each set numbers from least to greatest.

15) – 18, – 22, 28, – 17, 4   ___, ___, ___, ___, ___, ___

16) 19, –36, 11, – 12, 5   ___, ___, ___, ___, ___, ___

17) 27, – 56, 20, 1, – 27   ___, ___, ___, ___, ___, ___

18) 26, – 96, 2, – 26, 87, –75   ___, ___, ___, ___, ___, ___

19) –10, –71, 70, –26, –59, –39   ___, ___, ___, ___, ___, ___

20) 88, 4, 38, 7, 78, 9   ___, ___, ___, ___, ___, ___

21) 84, 14, 24, 0, 35, 22   ___, ___, ___, ___, ___, ___

Common Core Subject Test Mathematics Grade 5

Numbers in Word Form

✎ Write each number in words.

1) 372 _____

2) 605 _____

3) 550 _____

4) 351 _____

5) 793 _____

6) 647 _____

7) 3,219 _____

8) 5,326 _____

9) 2,842 _____

10) 4,691 _____

11) 5,531 _____

12) 7,360 _____

13) 2,532 _____

14) 8,014 _____

15) 11,242 _____

Common Core Subject Test Mathematics Grade 5

Roman Numerals

✍ Write in Romans numerals.

1	I	11	XI	21	XXI
2	II	12	XII	22	XXII
3	III	13	XIII	23	XXIII
4	IV	14	XIV	24	XXIV
5	V	15	XV	25	XXV
6	VI	16	XVI	26	XXVI
7	VII	17	XVII	27	XXVII
8	VIII	18	XVIII	28	XXVIII
9	IX	19	XIX	29	XXIX
10	X	20	XX	30	XXX

1) 11 _____   2) 21 _____

3) 24 _____   4) 16 _____

5) 27 _____   6) 29 _____

7) 12 _____   8) 28 _____

9) 15 _____   10) 20 _____

11) Add 16 + 14 and write in Roman numerals. _____

12) Subtract 34 – 5 and write in Roman numerals. _____

Common Core Subject Test Mathematics Grade 5

Rounding Numbers

🖎 Round each number to the underlined place value.

1) 3,<u>7</u>93

2) 3,<u>8</u>76

3) 3,4<u>5</u>2

4) 7,1<u>9</u>3

5) 5,2<u>7</u>8

6) 1,4<u>7</u>7

7) 8,<u>3</u>13

8) 24.<u>6</u>8

9) 8<u>4</u>.92

10) 71.<u>3</u>4

11) 66<u>4</u>.7

12) <u>9</u>,135

13) 15.3<u>8</u>1

14) 4,<u>5</u>21

15) 3<u>6</u>.50

16) 4,<u>8</u>19

17) 6,6<u>8</u>5

18) 2,5<u>3</u>8

19) 73.<u>6</u>2

20) 16,<u>5</u>27

21) 2<u>9</u>.720

22) 12,3<u>6</u>6

23) 31,<u>7</u>29

24) 7,8<u>3</u>8

Common Core Subject Test Mathematics Grade 5

Odd or Even

Identify whether each number is even or odd.

1) 18 _____

2) 27 _____

3) 21 _____

4) 17 _____

5) 67 _____

6) 76 _____

7) 80 _____

8) 53 _____

9) 58 _____

10) 98 _____

11) 49 _____

12) 113 _____

Circle the even number in each group.

13) 52, 11, 35, 73, 5, 29

14) 13, 15, 113, 87, 71, 18

15) 33, 45, 86, 59, 63, 87

16) 55, 32, 79, 51, 21, 83

Circle the odd number in each group.

17) 54, 36, 48, 76, 71, 100

18) 32, 56, 40, 74, 98, 67

19) 58, 92, 25, 78, 76, 50

20) 89, 12, 88, 42, 48, 120

Common Core Subject Test Mathematics Grade 5

Repeating Patterns

🖊 Circle the picture that comes next in each picture pattern.

1)

 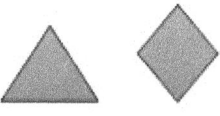

2)

3)

4)

5)

Growing Patterns

✏ Draw the picture that comes next in each growing pattern.

1)

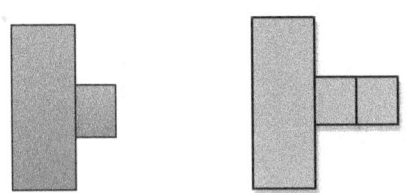

2)

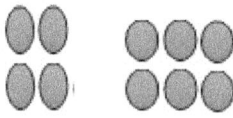

3)

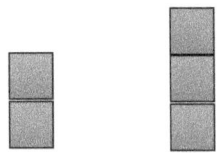

4)

5)

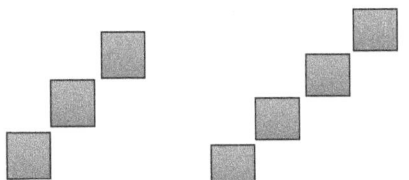

Common Core Subject Test Mathematics Grade 5

Patterns: Numbers

✎ Write the numbers that come next.

1) 2, 5, 8, 11,   ____, ____, ____, ____

2) 10, 15, 20, 25,   ____, ____, ____, ____

3) 4, 8, 12, 16,   ____, ____, ____, ____

4) 7, 17, 27, 37,   ____, ____, ____, ____

5) 5, 12, 19, 26,   ____, ____, ____, ____

6) 8, 16, 24, 32, 40,   ____, ____, ____, ____

✎ Write the next three numbers in each counting sequence.

1) −31, −19, −7, _____, _____, _____, _____

2) 541, 526, 511, _____, _____, _____, _____

3) 14, 34, _____, _____, 94, _____

4) 21, 29, _____, _____, _____

5) 89, 78, _____, _____, _____

6) 95, 82, 69, _____, _____, _____

7) 198, 166, 134, _____, _____, _____

8) What are the next three numbers in this counting sequence?

   1870, 1970, 2070, _____, _____, _____

9) What is the fourth number in this counting sequence?

   8, 14, 20, _____

WWW.MathNotion.Com

Common Core Subject Test Mathematics Grade 5

Answers of Worksheets

Place Values

1) 60 + 2
2) 50 + 6
3) 30 + 1
4) 40 + 5
5) 20 + 8
6) ones place
7) tens place
8) hundreds place
9) tens place
10) ones place

Comparing and Ordering Numbers

1) 31 less than 33
2) 57 greater than 49
3) 92 greater than 88
4) 76 greater than 67
5) 43 equals to 43
6) 54 greater than 46
7) 97 greater than 88
8) 42 greater than 36
9) 55 equals to 55
10) 57 less than 75
11) 28 less than 38
12) 19 greater than 15
13) 82 less than 90
14) 78 less than 84
15) –22, –18, –17, 4, 28
16) –36, –12, 5, 11, 19
17) –56, –27, 1, 20, 27
18) –96, –75, –26, 2, 26, 87
19) –71, –59, –39, –26, –10, 70
20) 4, 7, 9, 38, 78, 88
21) 0, 14, 22, 24, 35, 84

Numbers in Word Form

1) three hundred seventy-two
2) six hundred five
3) five hundred fifty
4) three hundred fifty-one
5) seven hundred ninety-three
6) six hundred forty-seven
7) three thousand, two hundred nineteen
8) five thousand, three hundred twenty-six
9) two thousand, eight hundred forty-two
10) four thousand, six hundred ninety-one
11) five thousand, five hundred thirty-one
12) seven thousand, three hundred sixty
13) two thousand, five hundred thirty-two
14) eight thousand, fourteen
15) eleven thousand, two hundred forty-two

Roman Numerals

1) XI
2) XXI
3) XXIV
4) XVI
5) XXVII
6) XXIX
7) XII
8) XXVIII
9) XV
10) XX
11) XXX
12) XXIX

Rounding Numbers

1) 4,000
2) 4,000
3) 3,450
4) 7,190
5) 5,280
6) 1,480
7) 8,300
8) 24.70
9) 85.00
10) 71.30
11) 665.00
12) 9,000

Common Core Subject Test Mathematics Grade 5

13) 15.380	16) 4,800	19) 73.60	22) 12,370
14) 4,500	17) 6,700	20) 16,500	23) 31,700
15) 37.00	18) 2,540	21) 30.00	24) 7,840

Odd or Even

1) even	6) even	11) odd	16) 32
2) odd	7) even	12) odd	17) 71
3) odd	8) odd	13) 52	18) 67
4) odd	9) even	14) 18	19) 25
5) odd	10) even	15) 86	20) 89

Repeating pattern

1)    2)    3)

4)    5)

Growing patterns

1) 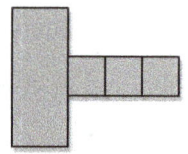   2) 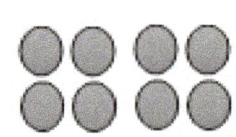   3)

4) 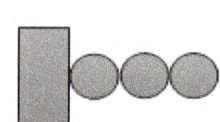   5)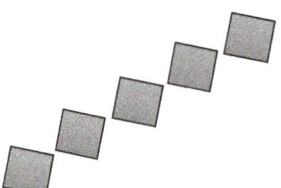

Patterns: Numbers

1) 2, 5, 8, 11, 14, 17, 20, 23   4) 7, 17, 27, 37, 47, 57, 67, 77

2) 10, 15, 20, 25, 30, 35, 40, 45   5) 5, 12, 19, 26, 33, 40, 47, 54

3) 4, 8, 12, 16, 20, 24, 28, 32   6) 8, 16, 24, 32, 40, 48, 56, 64

Patterns

1) 5, 17, 29, 41	4) 37, 45, 53	7) 102, 70, 38
2) 496, 481, 466, 451	5) 67, 56, 45	8) 2170, 2270, 2370
3) 14, 34, 54, 74, 94, 114	6) 56, 43, 30	9) 26

WWW.MathNotion.Com

Common Core Subject Test Mathematics Grade 5

Chapter 2 : Whole Number Operations

Topics that you'll learn in this chapter:

- ✓ Adding Whole Numbers,
- ✓ Subtracting Whole Numbers,
- ✓ Multiplying Whole Numbers,
- ✓ Dividing Hundreds,
- ✓ Long Division by One Digit,
- ✓ Division with Remainders,
- ✓ Rounding Whole Numbers,
- ✓ Whole Number Estimation,

Common Core Subject Test Mathematics Grade 5

Adding Whole Numbers

✎ Add.

1) 5,763
   + 8,238
   _____

2) 6,834
   + 4,998
   _____

3) 3,548
   + 5,693
   _____

4) 2,769
   + 8,872
   _____

5) 3,196
   + 2,936
   _____

6) 7,009
   + 4,992
   _____

✎ Find the missing numbers.

7) 3,468 + ___ = 4,102

8) 840 + 2,360 = ___

9) 5,200 + ___ = 7,980

10) 631 + ___ = 2,007

11) ___ + 803 = 3,945

12) ___ + 2,156 = 5,922

13) David sells gems. He finds a diamond in Istanbul and buys it for $4,795. Then, he flies to Cairo and purchases a bigger diamond for the bargain price of $9,633. How much does David spend on the two diamonds? _____

WWW.MathNotion.Com

Common Core Subject Test Mathematics Grade 5

Subtracting Whole Numbers

🖎 Subtract.

1) 10,512
   − 4,411
   _____

2) 5,204
   − 3,679
   _____

3) 8,520
   − 6,483
   _____

4) 8,001
   − 5,224
   _____

5) 11,916
   − 8,711
   _____

6) 5,005
   − 2,008
   _____

🖎 Find the missing number.

7) 5,263 − ___ = 2,367

8) 7,198 − ___ = 4,742

9) 8,928 − 3,764 = ___

10) 6,511 − ___ = 3,759

11) 7,003 − 5,489 = ___

12) 8,800 − 5,995 = ___

13) Jackson had $7,189 invested in the stock market until he lost $3,793 on those investments. How much money does he have in the stock market now?
_____

Common Core Subject Test Mathematics Grade 5

Multiplying Whole Numbers

✏ Find the answers.

1) 2,200
 × 31
 _____

2) 3,200
 × 22
 _____

3) 5,790
 × 5
 _____

4) 5,220
 × 3
 _____

5) 6,911
 × 3
 _____

6) 1,998
 ×40
 _____

7) 2,893
 ×5.5
 _____

8) 2,254
 × 3.5
 _____

9) 4,372
 × 4.8
 _____

10) 3,984
 × 2.75
 _____

11) 4,900
 × 2.5
 _____

12) 8,200
 × 4.5
 _____

Common Core Subject Test Mathematics Grade 5

Dividing Hundreds

✎ Find answers.

1) $4,440 \div 400$

2) $1,600 \div 40$

3) $9,990 \div 90$

4) $4,200 \div 60$

5) $6,400 \div 8,000$

6) $2,700 \div 30$

7) $3,333 \div 30$

8) $558 \div 45$

9) $2,278 \div 85$

10) $1,683 \div 55$

11) $1,582 \div 35$

12) $9,000 \div 600$

13) $1,000 \div 2,500$

14) $44.8 \div 20$

15) $6,800 \div 400$

16) $1,500 \div 5,000$

17) $36.60 \div 120$

18) $7,700 \div 700$

19) $5,400 \div 600$

20) $8,000 \div 160$

21) $18,000 \div 9,000$

22) $42,000 \div 30$

23) $480 \div 40$

24) $63,000 \div 900$

Common Core Subject Test Mathematics Grade 5

Long Division by Two Digits

✎ Find the quotient.

1) 18)576   10) 41)1,476

2) 14)952   11) 53)2,491

3) 21)588   12) 60)2,880

4) 23)299   13) 32)2,912

5) 44)748   14) 77)8,393

6) 26)234   15) 85)3,740

7) 16)496   16) 57)4,617

8) 29)1,479   17) 50)9,200

9) 54)1,080   18) 25)15,400

Division with Remainders

✎ Find the quotient with remainder.

1) 14)715   8) 65)8,624

2) 16)2,750   9) 35)5,705

3) 27)4,603   10) 92)13,161

4) 58)2,554   11) 46)12,214

5) 42)7,732   12) 69)42,482

6) 63)6,737   13) 85)6,858

7) 71)9,036   14) 87)34,304

WWW.MathNotion.Com

Common Core Subject Test Mathematics Grade 5

Rounding Whole Numbers

Round each number to the underlined place value.

1) 7,5̲33

2) 9,3̲74

3) 8,8̲3

4) 2,3̲68

5) 5,5̲7̲7

6) 3,3̲81

7) 3,5̲20

8) 9,3̲38

9) 8.5̲81

10) 33.5̲7

11) 51.6̲9

12) 22.1̲38

13) 6̲,758

14) 11,5̲57

15) 8,8̲3̲8

16) 5.8̲89

17) 1.8̲60

18) 25.0̲70

19) 9̲.332

20) 49.4̲8

21) 28.8̲9

22) 24,3̲7̲7

23) 52,1̲5̲8

24) 13,8̲3

25) 9,6̲09

26) 17,45̲1

27) 18,7̲68

Common Core Subject Test Mathematics Grade 5

Whole Number Estimation

✏️ Estimate the sum by rounding each added to the nearest ten.

1) 875 + 325

2) 985 + 1,452

3) 2,424 + 4,128

4) 1,576 + 6,279

5) 1,247 + 3,863

6) 6,746 + 5,121

7) 3,924 + 6,456

8) 1,785 + 7,164

9) 1,458
   + 2,442
   _____

10) 5,689
    + 4,151
    _____

11) 8,259
    + 4,754
    _____

12) 6,788
    + 3,954
    _____

13) 9,123
    + 4,455
    _____

14) 6,680
    + 5,358
    _____

15) 3,165
    + 7,124
    _____

16) 8,859
    + 6,452
    _____

WWW.MathNotion.Com

Common Core Subject Test Mathematics Grade 5

Answers of Worksheets

Adding Whole Numbers

1) 14,001   6) 12,001   11) 3,142
2) 11,832   7) 634   12) 3,766
3) 9,241   8) 3,200   13) $14,428
4) 11,641   9) 2,780
5) 6,132   10) 1,376

Subtracting Whole Numbers

1) 6,101   6) 2,997   11) 1,514
2) 1,525   7) 2,896   12) 2,805
3) 2,037   8) 2,456   13) 3,396
4) 2,777   9) 5,164
5) 3,205   10) 2,752

Multiplying Whole Numbers

1) 68,200   5) 20,733   9) 20,985.6
2) 70,400   6) 79,920   10) 10,956
3) 28,950   7) 15,911.5   11) 12,250
4) 15,660   8) 7,889   12) 36,900

Dividing Hundreds

1) 11.1   7) 111.1   13) 0.4   19) 9
2) 40   8) 12.4   14) 2.24   20) 50
3) 111   9) 26.8   15) 17   21) 2
4) 70   10) 30.6   16) 0.3   22) 1,400
5) 0.8   11) 45.2   17) 0.305   23) 12
6) 90   12) 15   18) 11   24) 70

Long Division by Two Digits

1) 32   6) 9   11) 47   16) 81
2) 68   7) 31   12) 48   17) 184
3) 28   8) 51   13) 91   18) 616
4) 13   9) 20   14) 109
5) 17   10) 36   15) 44

WWW.MathNotion.Com

Common Core Subject Test Mathematics Grade 5

Division with Remainders

1) 51 R1
2) 171 R14
3) 170 R13
4) 44 R2
5) 184 R4
6) 106 R59
7) 127 R19
8) 132 R44
9) 163 R0
10) 143 R5
11) 265 R24
12) 615 R47
13) 80 R58
14) 394 R26

Rounding Whole Numbers

1) 7,500
2) 9,400
3) 8,880
4) 2,370
5) 5,580
6) 3,380
7) 3,500
8) 9,340
9) 8.60
10) 33.60
11) 51.70
12) 22.100
13) 7,000
14) 11,560
15) 8,840
16) 5.900
17) 1.900
18) 25.100
19) 9.000
20) 49.50
21) 28.90
22) 24,380
23) 52,160
24) 13,880
25) 9,600
26) 17,450
27) 18,800

Whole Number Estimation

1) 1,200
2) 2,440
3) 6,550
4) 7,860
5) 5,110
6) 11,870
7) 10,380
8) 8,950
9) 3,900
10) 9,840
11) 13,010
12) 10,740
13) 13,580
14) 12,040
15) 10,290
16) 15,310

Chapter 3 : Number Theory

Topics that you'll learn in this chapter:

- ✓ Factoring Numbers,
- ✓ Prime Factorization,
- ✓ Divisibility Rules,
- ✓ Greatest Common Factor,
- ✓ Least Common Multiple,

Common Core Subject Test Mathematics Grade 5

Factoring Numbers

✎ List all positive factors of each number.

1) 12   6) 56   11) 27
2) 16   7) 65   12) 63
3) 28   8) 70   13) 72
4) 34   9) 25   14) 15
5) 95   10) 48   15) 80

✎ List the prime factorization for each number.

16) 10   19) 30   22) 55
17) 26   20) 40   23) 78
18) 20   21) 44   24) 96

Prime Factorization

✎ Factor the following numbers to their prime factors.

1) 6   9) 58   17) 69
2) 49   10) 62   18) 76
3) 60   11) 75   19) 86
4) 4   12) 88   20) 92
5) 46   13) 93   21) 99
6) 57   14) 100   22) 77
7) 54   15) 68   23) 90
8) 38   16) 90   24) 74

Divisibility Rules

✎ Use the divisibility rules to underline the factors of the number.

1) 8   2 3 4 5 6 7 8 9 10

2) 18   2 3 4 5 6 7 8 9 10

3) 55   2 3 4 5 6 7 8 9 10

4) 45   2 3 4 5 6 7 8 9 10

5) 20   2 3 4 5 6 7 8 9 10

6) 9   2 3 4 5 6 7 8 9 10

7) 21   2 3 4 5 6 7 8 9 10

8) 28   2 3 4 5 6 7 8 9 10

9) 36   2 3 4 5 6 7 8 9 10

10) 40   2 3 4 5 6 7 8 9 10

11) 39   2 3 4 5 6 7 8 9 10

12) 51   2 3 4 5 6 7 8 9 10

Common Core Subject Test Mathematics Grade 5

Greatest Common Factor

✎ Find the GCF for each number pair.

1) 25, 15   9) 52, 3   17) 66, 18

2) 8, 18   10) 12, 54   18) 70, 15

3) 14, 28   11) 11, 13   19) 38, 14

4) 18, 32   12) 56, 48   20) 36, 28

5) 15, 45   13) 75, 25   21) 100, 60

6) 22, 33   14) 40, 60   22) 85, 35

7) 19, 21   15) 52, 32   23) 16, 48

8) 27, 72   16) 30, 55   24) 13, 39

Least Common Multiple

✎ Find the LCM for each number pair.

1) 3, 15   9) 13, 26   17) 13, 2, 26

2) 5, 35   10) 15, 65   18) 18, 6, 24

3) 24, 16   11) 12, 8   19) 9, 12, 15

4) 28, 40   12) 6, 44   20) 7, 12, 4

5) 9, 27   13) 10, 16   21) 5, 15, 16

6) 46, 23   14) 7, 6   22) 13, 4, 26

7) 22, 66   15) 12, 36, 24   23) 3, 14, 5

8) 4, 9   16) 5, 11, 2   24) 32, 8, 3

WWW.MathNotion.Com

Common Core Subject Test Mathematics Grade 5

Answers of Worksheets

Factoring Numbers

1) 1, 2, 3, 4, 6, 12
2) 1, 2, 4, 8, 16
3) 1, 2, 4, 7, 14, 28
4) 1, 2, 17, 34
5) 1, 5, 19, 95
6) 1, 2, 4, 7, 8, 14, 28, 56
7) 1, 5, 13, 65
8) 1, 2, 5, 7, 10, 14, 35, 70
9) 1, 5, 25
10) 1, 2, 3, 4, 6, 8, 12, 16, 24, 48
11) 1, 3, 9, 27
12) 1, 3, 7, 9, 21, 63
13) 1, 2, 3, 4, 6, 8, 9, 12, 18, 24, 36, 72
14) 1, 3, 5, 15
15) 1, 2, 4, 5, 8, 10, 16, 20, 40, 80
16) 2×5
17) 2×13
18) $2 \times 2 \times 5$
19) $2 \times 3 \times 5$
20) $2 \times 2 \times 2 \times 5$
21) $2 \times 2 \times 11$
22) 5×11
23) $2 \times 3 \times 13$
24) $2 \times 2 \times 2 \times 2 \times 2 \times 3$

Prime Factorization

1) 2. 3
2) 7. 7
3) 2. 2. 3. 5
4) 2. 2
5) 2. 23
6) 3. 19
7) 2. 3. 3. 3
8) 2. 19
9) 2. 29
10) 2. 31
11) 3. 5. 5
12) 2. 2. 2. 11
13) 3. 31
14) 2. 2. 5. 5
15) 2. 2. 17
16) 2. 3. 3. 5
17) 3. 23
18) 2. 2. 19
19) 2. 43
20) 2. 2. 23
21) 3. 3. 11
22) 7. 11
23) 2. 3. 3. 5
24) 2. 37

Divisibility Rules

1) 8   <u>2</u>   3   <u>4</u>   5   6   7   <u>8</u>   9   10
2) 18   <u>2</u>   <u>3</u>   4   5   <u>6</u>   7   8   <u>9</u>   10
3) 55   2   3   4   <u>5</u>   6   7   8   9   10
4) 45   2   <u>3</u>   4   <u>5</u>   6   7   8   <u>9</u>   10
5) 20   <u>2</u>   3   <u>4</u>   <u>5</u>   6   7   8   9   <u>10</u>
6) 9   2   <u>3</u>   4   5   6   7   8   <u>9</u>   10

Common Core Subject Test Mathematics Grade 5

7) 21   2 <u>3</u> 4 5 6 <u>7</u> 8 9 10

8) 28   <u>2</u> 3 <u>4</u> 5 6 <u>7</u> 8 9 10

9) 36   <u>2</u> <u>3</u> <u>4</u> 5 <u>6</u> 7 8 <u>9</u> 10

10) 40   <u>2</u> 3 <u>4</u> <u>5</u> 6 7 <u>8</u> 9 <u>10</u>

11) 39   2 <u>3</u> 4 5 6 7 8 9 10

12) 51   2 <u>3</u> 4 5 6 7 8 9 10

Greatest Common Factor

1) 5	7) 1	13) 25	19) 2
2) 2	8) 9	14) 20	20) 4
3) 14	9) 1	15) 4	21) 20
4) 2	10) 6	16) 5	22) 5
5) 15	11) 1	17) 6	23) 16
6) 11	12) 8	18) 5	24) 13

Least Common Multiple

1) 15	7) 66	13) 80	19) 180
2) 35	8) 36	14) 42	20) 84
3) 48	9) 26	15) 72	21) 240
4) 280	10) 195	16) 110	22) 52
5) 27	11) 24	17) 26	23) 210
6) 46	12) 132	18) 72	24) 96

Common Core Subject Test Mathematics Grade 5

Chapter 4 : Fractions and Mixed Numbers

Topics that you'll learn in this chapter:

- ✓ Simplifying Fractions,
- ✓ Like Denominators,
- ✓ Compare Fractions with Like Denominators,
- ✓ More than two Fractions with Like Denominators,
- ✓ Unlike Denominators,
- ✓ Ordering Fractions,
- ✓ Denominators of 10, 100, and 1000,
- ✓ Fractions to Mixed Numbers,
- ✓ Mixed Numbers to Fractions,
- ✓ Add and Subtract Mixed Numbers,

Common Core Subject Test Mathematics Grade 5

Simplifying Fractions

✎ Simplify the fractions.

1) $\dfrac{44}{84}$

2) $\dfrac{8}{20}$

3) $\dfrac{12}{16}$

4) $\dfrac{4}{24}$

5) $\dfrac{15}{30}$

6) $\dfrac{9}{63}$

7) $\dfrac{4}{14}$

8) $\dfrac{17}{51}$

9) $\dfrac{24}{30}$

10) $\dfrac{5}{35}$

11) $\dfrac{16}{48}$

12) $\dfrac{33}{22}$

13) $\dfrac{45}{63}$

14) $\dfrac{2.4}{3.2}$

15) $\dfrac{12}{60}$

16) $\dfrac{70}{112}$

17) $\dfrac{2.7}{7.2}$

18) $\dfrac{33}{88}$

19) $\dfrac{1.5}{13.5}$

20) $\dfrac{39}{52}$

21) $\dfrac{5}{45}$

22) $\dfrac{2.1}{4.2}$

WWW.MathNotion.Com

Common Core Subject Test Mathematics Grade 5

Like Denominators

🖎 Add fractions.

1) $\dfrac{3}{4}+\dfrac{1}{4}$

2) $\dfrac{1}{5}+\dfrac{4}{5}$

3) $\dfrac{4}{9}+\dfrac{7}{9}$

4) $\dfrac{2}{7}+\dfrac{2}{7}$

5) $\dfrac{5}{13}+\dfrac{2}{13}$

6) $\dfrac{1}{14}+\dfrac{4}{14}$

7) $\dfrac{11}{19}+\dfrac{1}{19}$

8) $\dfrac{3}{16}+\dfrac{9}{16}$

9) $\dfrac{3}{10}+\dfrac{1}{10}$

10) $\dfrac{6}{17}+\dfrac{2}{17}$

11) $\dfrac{5}{22}+\dfrac{5}{22}$

12) $\dfrac{7}{35}+\dfrac{11}{35}$

13) $\dfrac{7}{27}+\dfrac{20}{27}$

14) $\dfrac{2}{31}+\dfrac{10}{31}$

15) $\dfrac{5}{23}+\dfrac{3}{23}$

16) $\dfrac{8}{41}+\dfrac{13}{41}$

17) $\dfrac{15}{37}+\dfrac{18}{37}$

18) $\dfrac{2}{51}+\dfrac{7}{51}$

19) $\dfrac{17}{26}+\dfrac{6}{26}$

20) $\dfrac{12}{48}+\dfrac{11}{48}$

21) $\dfrac{11}{29}+\dfrac{8}{29}$

22) $\dfrac{15}{34}+\dfrac{19}{34}$

23) $\dfrac{1}{19}+\dfrac{5}{19}$

24) $\dfrac{3}{53}+\dfrac{4}{53}$

25) $\dfrac{3}{20}+\dfrac{6}{20}$

26) $\dfrac{2}{63}+\dfrac{6}{63}$

27) $\dfrac{6}{38}+\dfrac{1}{38}$

28) $\dfrac{14}{31}+\dfrac{17}{31}$

29) $\dfrac{3}{28}+\dfrac{5}{28}$

30) $\dfrac{2}{37}+\dfrac{15}{37}$

WWW.MathNotion.Com

Common Core Subject Test Mathematics Grade 5

➤ **Subtract fractions.**

1) $\dfrac{8}{9} - \dfrac{4}{9}$

2) $\dfrac{3}{8} - \dfrac{1}{8}$

3) $\dfrac{9}{11} - \dfrac{3}{11}$

4) $\dfrac{9}{14} - \dfrac{4}{14}$

5) $\dfrac{15}{20} - \dfrac{8}{20}$

6) $\dfrac{8}{15} - \dfrac{7}{15}$

7) $\dfrac{11}{19} - \dfrac{9}{19}$

8) $\dfrac{13}{16} - \dfrac{1}{16}$

9) $\dfrac{7}{29} - \dfrac{4}{29}$

10) $\dfrac{14}{23} - \dfrac{7}{23}$

11) $\dfrac{15}{34} - \dfrac{7}{34}$

12) $\dfrac{18}{41} - \dfrac{9}{41}$

13) $\dfrac{17}{39} - \dfrac{16}{39}$

14) $\dfrac{6}{26} - \dfrac{2}{26}$

15) $\dfrac{14}{17} - \dfrac{4}{17}$

16) $\dfrac{33}{55} - \dfrac{20}{55}$

17) $\dfrac{41}{49} - \dfrac{36}{49}$

18) $\dfrac{40}{53} - \dfrac{39}{53}$

19) $\dfrac{27}{37} - \dfrac{17}{37}$

20) $\dfrac{21}{47} - \dfrac{11}{47}$

21) $\dfrac{24}{43} - \dfrac{12}{43}$

22) $\dfrac{13}{19} - \dfrac{12}{19}$

23) $\dfrac{6}{26} - \dfrac{3}{26}$

24) $\dfrac{9}{15} - \dfrac{7}{15}$

25) $\dfrac{8}{39} - \dfrac{3}{39}$

26) $\dfrac{18}{61} - \dfrac{15}{61}$

27) $\dfrac{12}{53} - \dfrac{9}{53}$

28) $\dfrac{75}{76} - \dfrac{74}{76}$

29) $\dfrac{26}{45} - \dfrac{13}{45}$

30) $\dfrac{20}{57} - \dfrac{17}{57}$

Common Core Subject Test Mathematics Grade 5

Compare Fractions with Like Denominators

✎ Evaluate and compare. Write < or > or =.

1) $\frac{1}{3} + \frac{1}{3} \text{---} \frac{1}{3}$

2) $\frac{3}{6} + \frac{3}{6} \text{---} \frac{5}{6}$

3) $\frac{8}{9} - \frac{4}{9} \text{---} \frac{7}{9}$

4) $\frac{4}{11} + \frac{5}{11} \text{---} \frac{7}{11}$

5) $\frac{9}{14} - \frac{8}{14} \text{---} \frac{5}{14}$

6) $\frac{11}{17} - \frac{3}{17} \text{---} \frac{6}{17}$

7) $\frac{11}{21} + \frac{2}{21} \text{---} \frac{10}{21}$

8) $\frac{8}{32} + \frac{6}{32} \text{---} \frac{9}{32}$

9) $\frac{25}{29} - \frac{16}{29} \text{---} \frac{11}{29}$

10) $\frac{28}{41} + \frac{13}{41} \text{---} \frac{27}{41}$

11) $\frac{18}{35} - \frac{11}{35} \text{---} \frac{22}{35}$

12) $\frac{32}{47} - \frac{22}{47} \text{---} \frac{11}{47}$

13) $\frac{14}{27} + \frac{13}{27} \text{---} \frac{24}{27}$

14) $\frac{34}{52} - \frac{11}{52} \text{---} \frac{21}{52}$

15) $\frac{43}{56} - \frac{24}{56} \text{---} \frac{27}{56}$

16) $\frac{27}{71} + \frac{25}{71} \text{---} \frac{48}{71}$

Common Core Subject Test Mathematics Grade 5

More Than Two Fractions with Like Denominators

✏️ **Add fractions.**

1) $\dfrac{5}{9} + \dfrac{2}{9} + \dfrac{2}{9}$

2) $\dfrac{4}{6} + \dfrac{1}{6} + \dfrac{1}{6}$

3) $\dfrac{2}{17} + \dfrac{4}{17} + \dfrac{2}{17}$

4) $\dfrac{1}{5} + \dfrac{1}{5} + \dfrac{1}{5}$

5) $\dfrac{7}{18} + \dfrac{2}{18} + \dfrac{3}{18}$

6) $\dfrac{3}{27} + \dfrac{5}{27} + \dfrac{2}{27}$

7) $\dfrac{4}{33} + \dfrac{4}{33} + \dfrac{4}{33}$

8) $\dfrac{8}{23} + \dfrac{6}{23} + \dfrac{2}{23}$

9) $\dfrac{13}{41} + \dfrac{2}{41} + \dfrac{8}{41}$

10) $\dfrac{6}{35} + \dfrac{9}{35} + \dfrac{20}{35}$

11) $\dfrac{1}{37} + \dfrac{5}{37} + \dfrac{5}{37}$

12) $\dfrac{4}{43} + \dfrac{9}{43} + \dfrac{8}{43}$

13) $\dfrac{4}{51} + \dfrac{10}{51} + \dfrac{7}{51}$

14) $\dfrac{5}{26} + \dfrac{13}{26} + \dfrac{6}{26}$

15) $\dfrac{5}{64} + \dfrac{4}{64} + \dfrac{2}{64}$

16) $\dfrac{1}{73} + \dfrac{5}{73} + \dfrac{6}{73}$

WWW.MathNotion.Com

Unlike Denominators

🖉 Add fraction.

1) $\dfrac{2}{9} + \dfrac{3}{4}$

2) $\dfrac{1}{4} + \dfrac{3}{5}$

3) $\dfrac{1}{16} + \dfrac{3}{4}$

4) $\dfrac{3}{8} + \dfrac{1}{7}$

5) $\dfrac{1}{3} + \dfrac{2}{4}$

6) $\dfrac{1}{6} + \dfrac{3}{7}$

7) $\dfrac{5}{18} + \dfrac{4}{6}$

8) $\dfrac{1}{12} + \dfrac{5}{6}$

9) $\dfrac{5}{27} + \dfrac{1}{9}$

10) $\dfrac{1}{6} + \dfrac{7}{24}$

11) $\dfrac{3}{5} + \dfrac{1}{8}$

12) $\dfrac{11}{42} + \dfrac{3}{7}$

13) $\dfrac{7}{20} + \dfrac{1}{3}$

14) $\dfrac{1}{45} + \dfrac{3}{5}$

15) $\dfrac{3}{32} + \dfrac{5}{8}$

16) $\dfrac{3}{48} + \dfrac{5}{6}$

17) $\dfrac{5}{12} + \dfrac{1}{6}$

18) $\dfrac{1}{34} + \dfrac{3}{17}$

19) $\dfrac{4}{9} + \dfrac{7}{54}$

20) $\dfrac{13}{56} + \dfrac{4}{7}$

21) $\dfrac{3}{12} + \dfrac{2}{3}$

22) $\dfrac{4}{33} + \dfrac{5}{11}$

Common Core Subject Test Mathematics Grade 5

✏️ **Subtract fractions.**

1) $\dfrac{8}{9} - \dfrac{1}{2}$

2) $\dfrac{2}{3} - \dfrac{3}{10}$

3) $\dfrac{1}{6} - \dfrac{1}{9}$

4) $\dfrac{7}{8} - \dfrac{1}{4}$

5) $\dfrac{3}{4} - \dfrac{1}{28}$

6) $\dfrac{11}{30} - \dfrac{3}{15}$

7) $\dfrac{11}{18} - \dfrac{5}{9}$

8) $\dfrac{5}{13} - \dfrac{3}{26}$

9) $\dfrac{17}{35} - \dfrac{2}{7}$

10) $\dfrac{5}{6} - \dfrac{12}{36}$

11) $\dfrac{5}{9} - \dfrac{1}{27}$

12) $\dfrac{3}{5} - \dfrac{1}{8}$

13) $\dfrac{2}{3} - \dfrac{3}{5}$

14) $\dfrac{7}{8} - \dfrac{3}{7}$

15) $\dfrac{5}{9} - \dfrac{13}{45}$

16) $\dfrac{3}{4} - \dfrac{5}{36}$

17) $\dfrac{39}{49} - \dfrac{5}{7}$

18) $\dfrac{3}{11} - \dfrac{3}{22}$

19) $\dfrac{17}{48} - \dfrac{4}{12}$

20) $\dfrac{2}{3} - \dfrac{4}{13}$

21) $\dfrac{5}{8} - \dfrac{19}{72}$

22) $\dfrac{3}{5} - \dfrac{1}{12}$

Ordering Fractions

Order the fractions from least to greatest.

1) $\frac{1}{5}, \frac{1}{11}, \frac{1}{8}, \frac{1}{3}$ ____, ____, ____, ____

2) $\frac{1}{9}, \frac{1}{18}, \frac{2}{4}, \frac{1}{5}$ ____, ____, ____, ____

3) $\frac{4}{7}, \frac{1}{7}, \frac{6}{21}, \frac{15}{21}$ ____, ____, ____, ____

4) $\frac{1}{2}, \frac{1}{3}, \frac{4}{9}, \frac{5}{18}$ ____, ____, ____, ____

5) $\frac{4}{9}, \frac{3}{4}, \frac{7}{36}, \frac{1}{6}$ ____, ____, ____, ____

Order the fractions from greatest to least.

6) $\frac{3}{4}, \frac{4}{7}, \frac{3}{10}, \frac{5}{13}$ ____, ____, ____, ____

7) $\frac{5}{11}, \frac{5}{6}, \frac{2}{5}, \frac{1}{3}$ ____, ____, ____, ____

8) $\frac{7}{8}, \frac{1}{6}, \frac{3}{4}, \frac{5}{15}$ ____, ____, ____, ____

9) $\frac{4}{7}, \frac{2}{3}, \frac{11}{25}, \frac{13}{33}$ ____, ____, ____, ____

10) $\frac{18}{20}, \frac{15}{16}, \frac{14}{18}, \frac{5}{12}$ ____, ____, ____, ____

Denominators of 10, 100, and 1000

✎ Add fractions.

1) $\dfrac{7}{10}+\dfrac{13}{100}$

2) $\dfrac{1}{10}+\dfrac{10}{100}$

3) $\dfrac{15}{100}+\dfrac{1}{1,000}$

4) $\dfrac{56}{100}+\dfrac{3}{10}$

5) $\dfrac{50}{1,000}+\dfrac{7}{10}$

6) $\dfrac{6}{10}+\dfrac{30}{1,000}$

7) $\dfrac{9}{100}+\dfrac{3}{10}$

8) $\dfrac{5}{10}+\dfrac{50}{100}$

9) $\dfrac{48}{100}+\dfrac{6}{10}$

10) $\dfrac{70}{100}+\dfrac{2}{10}$

11) $\dfrac{80}{100}+\dfrac{200}{1,000}$

12) $\dfrac{30}{100}+\dfrac{4}{10}$

13) $\dfrac{9}{100}+\dfrac{7}{10}$

14) $\dfrac{25}{100}+\dfrac{6}{10}$

15) $\dfrac{15}{100}+\dfrac{8}{10}$

16) $\dfrac{3}{10}+\dfrac{31}{100}$

17) $\dfrac{8}{10}+\dfrac{11}{100}$

18) $\dfrac{34}{100}+\dfrac{6}{10}$

Common Core Subject Test Mathematics Grade 5

✏ Subtract fractions.

1) $\dfrac{8}{10} - \dfrac{20}{100}$

2) $\dfrac{5}{10} - \dfrac{47}{100}$

3) $\dfrac{12}{100} - \dfrac{60}{1,000}$

4) $\dfrac{6}{10} - \dfrac{50}{100}$

5) $\dfrac{3}{10} - \dfrac{23}{100}$

6) $\dfrac{70}{100} - \dfrac{250}{1,000}$

7) $\dfrac{4}{10} - \dfrac{350}{1,000}$

8) $\dfrac{70}{100} - \dfrac{3}{10}$

9) $\dfrac{40}{100} - \dfrac{3}{10}$

10) $\dfrac{6}{10} - \dfrac{180}{1,000}$

11) $\dfrac{93}{100} - \dfrac{5}{10}$

12) $\dfrac{65}{100} - \dfrac{4}{10}$

13) $\dfrac{80}{100} - \dfrac{6}{10}$

14) $\dfrac{90}{100} - \dfrac{5}{10}$

15) $\dfrac{200}{1,000} - \dfrac{1}{10}$

16) $\dfrac{90}{100} - \dfrac{7}{10}$

17) $\dfrac{900}{1,000} - \dfrac{40}{100}$

18) $\dfrac{60}{100} - \dfrac{3}{10}$

Common Core Subject Test Mathematics Grade 5

Fractions to Mixed Numbers

✏ Convert fractions to mixed numbers.

1) $\dfrac{9}{5}$

2) $\dfrac{11}{3}$

3) $\dfrac{39}{8}$

4) $\dfrac{27}{11}$

5) $\dfrac{7}{2}$

6) $\dfrac{43}{4}$

7) $\dfrac{49}{9}$

8) $\dfrac{15}{4}$

9) $\dfrac{37}{7}$

10) $\dfrac{19}{7}$

11) $\dfrac{41}{9}$

12) $\dfrac{45}{12}$

13) $\dfrac{17}{5}$

14) $\dfrac{29}{6}$

15) $\dfrac{13}{4}$

16) $\dfrac{15}{7}$

17) $\dfrac{65}{7}$

18) $\dfrac{59}{8}$

19) $\dfrac{25}{4}$

20) $\dfrac{17}{8}$

WWW.MathNotion.Com

Common Core Subject Test Mathematics Grade 5

Mixed Numbers to Fractions

✎ Convert to fraction.

1) $3\frac{3}{5}$

2) $1\frac{1}{3}$

3) $4\frac{2}{5}$

4) $4\frac{2}{8}$

5) $2\frac{1}{5}$

6) $2\frac{8}{11}$

7) $4\frac{4}{7}$

8) $3\frac{7}{12}$

9) $2\frac{1}{3}$

10) $7\frac{5}{7}$

11) $2\frac{7}{10}$

12) $3\frac{4}{9}$

13) $1\frac{5}{8}$

14) $4\frac{3}{11}$

15) $3\frac{4}{7}$

16) $5\frac{2}{8}$

17) $7\frac{1}{7}$

18) $13\frac{1}{2}$

19) $4\frac{2}{7}$

20) $5\frac{2}{10}$

21) $12\frac{1}{3}$

22) $7\frac{1}{8}$

Common Core Subject Test Mathematics Grade 5

Add and Subtract Mixed Numbers

✎ Add mixed numbers.

1) $3\frac{2}{5} + 8\frac{1}{5}$

2) $3\frac{2}{3} + 4\frac{1}{2}$

3) $6\frac{2}{7} + 2\frac{3}{7}$

4) $4\frac{2}{5} + 3\frac{1}{4}$

5) $8\frac{3}{4} - 2\frac{1}{2}$

6) $6\frac{5}{12} - 4\frac{1}{4}$

7) $5\frac{3}{8} - 3\frac{7}{8}$

8) $6\frac{1}{4} - 2\frac{15}{16}$

9) $9\frac{23}{28} - 4\frac{17}{28}$

10) $6\frac{1}{6} + 6\frac{2}{3}$

11) $4\frac{2}{9} + 5\frac{5}{9}$

12) $2\frac{1}{4} + 7\frac{4}{7}$

13) $7\frac{1}{5} - 3\frac{3}{5}$

14) $3\frac{1}{6} + 2\frac{3}{7}$

15) $2\frac{1}{3} + 4\frac{1}{4}$

16) $4\frac{1}{4} - 1\frac{2}{5}$

17) $\frac{1}{3} + 6\frac{1}{6}$

18) $2\frac{3}{5} + 2\frac{1}{10}$

Common Core Subject Test Mathematics Grade 5

Answers of Worksheets

Simplifying Fractions

1) $\frac{11}{21}$
2) $\frac{2}{5}$
3) $\frac{3}{4}$
4) $\frac{1}{6}$
5) $\frac{1}{2}$
6) $\frac{1}{7}$
7) $\frac{2}{7}$
8) $\frac{1}{3}$
9) $\frac{4}{5}$
10) $\frac{1}{7}$
11) $\frac{1}{3}$
12) $\frac{3}{2}$
13) $\frac{5}{7}$
14) $\frac{3}{4}$
15) $\frac{1}{5}$
16) $\frac{5}{8}$
17) $\frac{3}{8}$
18) $\frac{3}{8}$
19) $\frac{1}{9}$
20) $\frac{3}{4}$
21) $\frac{1}{9}$
22) $\frac{1}{2}$

Like Denominators (addition)

1) 1
2) 1
3) $\frac{11}{9}$
4) $\frac{4}{7}$
5) $\frac{7}{13}$
6) $\frac{5}{14}$
7) $\frac{12}{19}$
8) $\frac{3}{4}$
9) $\frac{2}{5}$
10) $\frac{8}{17}$
11) $\frac{5}{11}$
12) $\frac{18}{35}$
13) 1
14) $\frac{12}{31}$
15) $\frac{8}{23}$
16) $\frac{21}{41}$
17) $\frac{33}{37}$
18) $\frac{3}{17}$
19) $\frac{23}{26}$
20) $\frac{23}{48}$
21) $\frac{19}{29}$
22) 1
23) $\frac{6}{19}$
24) $\frac{7}{53}$
25) $\frac{9}{20}$
26) $\frac{8}{63}$
27) $\frac{7}{38}$
28) 1
29) $\frac{2}{7}$
30) $\frac{17}{37}$

Like Denominators (Subtraction)

1) $\frac{4}{9}$
2) $\frac{1}{4}$
3) $\frac{6}{11}$
4) $\frac{5}{14}$
5) $\frac{7}{20}$
6) $\frac{1}{15}$
7) $\frac{2}{19}$
8) $\frac{3}{4}$
9) $\frac{3}{29}$
10) $\frac{7}{23}$
11) $\frac{8}{34}$
12) $\frac{9}{41}$
13) $\frac{1}{39}$
14) $\frac{2}{13}$
15) $\frac{10}{17}$
16) $\frac{13}{55}$
17) $\frac{5}{49}$
18) $\frac{1}{53}$
19) $\frac{10}{37}$
20) $\frac{10}{47}$
21) $\frac{12}{43}$
22) $\frac{1}{19}$
23) $\frac{3}{26}$
24) $\frac{2}{15}$

Common Core Subject Test Mathematics Grade 5

25) $\frac{5}{39}$ 27) $\frac{3}{53}$ 29) $\frac{13}{45}$

26) $\frac{3}{61}$ 28) $\frac{1}{76}$ 30) $\frac{1}{19}$

Compare Fractions with Like Denominators

1) $\frac{2}{3} > \frac{1}{3}$ 5) $\frac{1}{14} < \frac{5}{14}$ 9) $\frac{9}{29} < \frac{11}{29}$ 13) $1 > \frac{24}{27}$

2) $1 > \frac{5}{6}$ 6) $\frac{8}{17} > \frac{6}{17}$ 10) $1 > \frac{27}{41}$ 14) $\frac{23}{52} > \frac{21}{52}$

3) $\frac{4}{9} < \frac{7}{9}$ 7) $\frac{13}{21} > \frac{10}{21}$ 11) $\frac{7}{35} < \frac{22}{35}$ 15) $\frac{19}{56} < \frac{27}{56}$

4) $\frac{9}{11} > \frac{7}{11}$ 8) $\frac{14}{32} > \frac{9}{32}$ 12) $\frac{10}{47} < \frac{11}{47}$ 16) $\frac{52}{71} > \frac{48}{71}$

More Than Two Fractions with Like Denominators

1) 1   5) $\frac{2}{3}$ 9) $\frac{23}{41}$ 13) $\frac{7}{17}$

2) 1   6) $\frac{10}{27}$ 10) 1   14) $\frac{12}{13}$

3) $\frac{8}{17}$ 7) $\frac{4}{11}$ 11) $\frac{11}{37}$ 15) $\frac{11}{64}$

4) $\frac{3}{5}$ 8) $\frac{16}{23}$ 12) $\frac{21}{43}$ 16) $\frac{12}{73}$

Unlike Denominators (Addition)

1) $\frac{35}{36}$ 7) $\frac{17}{18}$ 13) $\frac{41}{60}$ 19) $\frac{31}{54}$

2) $\frac{17}{20}$ 8) $\frac{11}{12}$ 14) $\frac{28}{45}$ 20) $\frac{45}{56}$

3) $\frac{13}{16}$ 9) $\frac{8}{27}$ 15) $\frac{23}{32}$ 21) $\frac{11}{12}$

4) $\frac{29}{56}$ 10) $\frac{11}{24}$ 16) $\frac{43}{48}$ 22) $\frac{19}{33}$

5) $\frac{5}{6}$ 11) $\frac{29}{40}$ 17) $\frac{7}{12}$

6) $\frac{25}{42}$ 12) $\frac{29}{42}$ 18) $\frac{7}{34}$

Unlike Denominators (Subtraction)

1) $\frac{7}{18}$ 5) $\frac{5}{7}$ 9) $\frac{1}{5}$ 13) $\frac{1}{15}$

2) $\frac{11}{30}$ 6) $\frac{1}{6}$ 10) $\frac{1}{2}$ 14) $\frac{25}{56}$

3) $\frac{1}{18}$ 7) $\frac{1}{18}$ 11) $\frac{14}{27}$ 15) $\frac{4}{15}$

4) $\frac{5}{8}$ 8) $\frac{7}{26}$ 12) $\frac{19}{40}$ 16) $\frac{11}{18}$

WWW.MathNotion.Com

Common Core Subject Test Mathematics Grade 5

17) $\frac{4}{49}$

18) $\frac{3}{22}$

19) $\frac{1}{48}$

20) $\frac{14}{39}$

21) $\frac{13}{36}$

22) $\frac{31}{60}$

Ordering Fractions

1) $\frac{1}{11}, \frac{1}{8}, \frac{1}{5}, \frac{1}{3}$

2) $\frac{1}{18}, \frac{1}{9}, \frac{1}{5}, \frac{2}{4}$,

3) $\frac{1}{7}, \frac{6}{21}, \frac{4}{7}, \frac{15}{21}$

4) $\frac{5}{18}, \frac{1}{3}, \frac{4}{9}, \frac{1}{2}$

5) $\frac{1}{6}, \frac{7}{36}, \frac{4}{9}, \frac{3}{4}$

6) $\frac{3}{4}, \frac{4}{7}, \frac{5}{13}, \frac{3}{10}$

7) $\frac{5}{6}, \frac{5}{11}, \frac{2}{5}, \frac{1}{3}$

8) $\frac{7}{8}, \frac{3}{4}, \frac{5}{15}, \frac{1}{6}$

9) $\frac{2}{3}, \frac{4}{7}, \frac{11}{25}, \frac{13}{33}$

10) $\frac{15}{16}, \frac{18}{20}, \frac{14}{18}, \frac{5}{12}$

Denominators of 10, 100, and 1000

1) $\frac{83}{100}$

2) $\frac{1}{5}$

3) $\frac{151}{1,000}$

4) $\frac{43}{50}$

5) $\frac{3}{4}$

6) $\frac{63}{100}$

7) $\frac{39}{100}$

8) 1

9) $\frac{27}{25}$

10) $\frac{9}{10}$

11) 1

12) $\frac{7}{10}$

13) $\frac{79}{100}$

14) $\frac{17}{20}$

15) $\frac{19}{20}$

16) $\frac{61}{100}$

17) $\frac{91}{100}$

18) $\frac{47}{50}$

Denominators of 10, 100, and 1000 (Subtract)

1) $\frac{3}{5}$

2) $\frac{3}{100}$

3) $\frac{3}{50}$

4) $\frac{1}{10}$

5) $\frac{7}{100}$

6) $\frac{9}{20}$

7) $\frac{1}{20}$

8) $\frac{2}{5}$

9) $\frac{1}{10}$

10) $\frac{21}{50}$

11) $\frac{43}{100}$

12) $\frac{1}{4}$

13) $\frac{1}{5}$

14) $\frac{2}{5}$

15) $\frac{1}{10}$

16) $\frac{1}{5}$

17) $\frac{1}{2}$

18) $\frac{3}{10}$

Fractions to Mixed Numbers

1) $1\frac{4}{5}$

2) $3\frac{2}{3}$

3) $4\frac{7}{8}$

4) $2\frac{5}{11}$

5) $3\frac{1}{2}$

6) $10\frac{3}{4}$

7) $5\frac{4}{9}$

8) $3\frac{3}{4}$

9) $5\frac{2}{7}$

10) $2\frac{5}{7}$

11) $4\frac{5}{9}$

12) $3\frac{9}{12}$

13) $3\frac{2}{5}$

14) $4\frac{5}{6}$

15) $3\frac{1}{4}$

16) $2\frac{1}{7}$

Common Core Subject Test Mathematics Grade 5

17) $9\frac{2}{7}$ 18) $7\frac{3}{8}$ 19) $6\frac{1}{4}$ 20) $2\frac{1}{8}$

Mixed Numbers to Fractions

1) $\frac{18}{5}$ 7) $\frac{32}{7}$ 13) $\frac{13}{8}$ 19) $\frac{30}{7}$

2) $\frac{4}{3}$ 8) $\frac{43}{12}$ 14) $\frac{47}{11}$ 20) $\frac{52}{10}$

3) $\frac{22}{5}$ 9) $\frac{7}{3}$ 15) $\frac{25}{7}$ 21) $\frac{37}{3}$

4) $\frac{34}{8}$ 10) $\frac{54}{7}$ 16) $\frac{42}{8}$ 22) $\frac{57}{8}$

5) $\frac{11}{5}$ 11) $\frac{27}{10}$ 17) $\frac{50}{7}$

6) $\frac{30}{11}$ 12) $\frac{31}{9}$ 18) $\frac{27}{2}$

Add and Subtract Mixed Numbers

1) $11\frac{3}{5}$ 6) $2\frac{1}{6}$ 11) $9\frac{7}{9}$ 16) $2\frac{17}{20}$

2) $8\frac{1}{6}$ 7) $1\frac{1}{2}$ 12) $9\frac{23}{28}$ 17) $6\frac{1}{2}$

3) $8\frac{5}{7}$ 8) $3\frac{5}{16}$ 13) $3\frac{3}{5}$ 18) $4\frac{7}{10}$

4) $7\frac{13}{20}$ 9) $5\frac{3}{14}$ 14) $5\frac{25}{42}$

5) $6\frac{1}{4}$ 10) $12\frac{5}{6}$ 15) $6\frac{7}{12}$

Chapter 5 : Decimals

Topics that you'll learn in this chapter:

- ✓ Adding and Subtracting Decimals,
- ✓ Multiplying and Dividing Decimals,
- ✓ Round decimals,
- ✓ Comparing Decimals,

Common Core Subject Test Mathematics Grade 5

Adding and Subtracting Decimals

✎ Add and subtract decimals.

1) 24.19 − 15.42

2) 42.23 + 25.42

3) 72.54 + 11.28

4) 57.45 − 24.75

5) 43.57 + 54.85

6) 86.68 − 54.12

✎ Solve.

7) ____ + 2.7 = 8.1

8) 6.4 + ____ = 12.8

9) 7.9 + ____ = 17

10) 4.6 + ____ = 15.3

11) ____ + 9.4 = 15

12) ____ + 8.24 = 13.54

✎ Order each set of numbers from least to greatest.

1) 0.4, 0.67, 0.44, 0.73, 0.51 ___, ___, ___, ___, ___, ___

2) 3.9, 6.1, 4.28, 7.02, 4.65 ___, ___, ___, ___, ___, ___

3) 1.9, 1.04, 0.79, 0.72, 0.09 ___, ___, ___, ___, ___, ___

4) 2.6, 5.2, 1.9, 4.01, 1.99, 3.2 ___, ___, ___, ___, ___, ___

5) 4.2, 6.1, 3.8, 5.7, 2.1, 2.8 ___, ___, ___, ___, ___, ___

6) 0.56, 0.87, 0.14, 1.24, 3.1 ___, ___, ___, ___, ___, ___

Multiplying and Dividing Decimals

✎ Find each product.

1) 1.5 × 2.1

2) 4.6 × 3.4

3) 6.3 × 2.5

4) 6.5 × 0.99

5) 12.1 × 5.2

6) 3.4 × 8.9

7) 4.8 × 9.1

8) 22.35 × 20

9) 15.25 × 3.6

✎ Find each quotient.

10) 3.5 ÷ 0.85

11) 15.35 ÷ 4.6

12) 32.42 ÷ 8.8

13) 9.2 ÷ 3.4

14) 0.84 ÷ 0.1

15) 21.5 ÷ 1,000

16) 4.1 ÷ 100

17) 9.7 ÷ 10

18) 6.55 ÷ 1.25

19) 18.48 ÷ 11.2

Rounding Decimals

Round each decimal number to the nearest place indicated.

1) 0.3<u>2</u>

2) 5.<u>0</u>1

3) 8.<u>8</u>24

4) 0.<u>4</u>78

5) <u>7</u>.32

6) 0.<u>2</u>9

7) 11.<u>3</u>1

8) <u>6</u>.223

9) 9.6<u>3</u>7

10) 5.<u>4</u>804

11) <u>7</u>.9

12) <u>5</u>.2439

13) 6.<u>4</u>92

14) 1.<u>6</u>2

15) 7<u>2</u>.85

16) 8<u>3</u>.67

17) 41.<u>6</u>8

18) 79<u>4</u>.741

19) 5<u>2</u>.2

20) 7<u>6</u>.93

21) <u>3</u>.219

22) 7<u>2</u>.09

23) 486.<u>4</u>91

24) 7.<u>0</u>8

Comparing Decimals

Write the correct comparison symbol (>, < or =).

1) 0.35 __ 1.8

2) 1.9 __ 1.19

3) 8.6 __ 8.6

4) 2.45 __ 24.5

5) 7.56 __ 0.756

6) 11.4 __ 11.05

7) 7.4 __ 0.7.4

8) 8.56 __ 0.85

9) 7 __ 0.7

10) 7.12 __ 0.712

11) 12.3 __ 12.5

12) 4.67 __ 4.68

13) 2.57 __ 2.75

14) 3.46 __ 0.346

15) 6.87 __ 6.78

16) 0.89 __ 0.98

17) 1.57 __ 0.157

18) 0.092 __ 0.091

19) 24.3 __ 24.3

20) 0.17 __ 0.71

21) 0.46 __ 0.64

22) 0.2 __ 0.08

23) 0.10 __ 0.1

24) 3.52 __ 31.5

Common Core Subject Test Mathematics Grade 5

Answers of Worksheets

Adding and Subtracting Decimals

1) 8.77	4) 32.7	7) 5.4	10) 10.7
2) 67.65	5) 98.42	8) 6.4	11) 5.6
3) 83.82	6) 32.56	9) 9.1	12) 5.3

Order and Comparing Decimals

1) 0.4, 0.44, 0.51, 0.67, 0.73
2) 3.9, 4.28, 4.65, 6.1, 7.02
3) 0.09, 0.72, 0.79, 1.04, 1.9
4) 1.9, 1.99, 2.6, 3.2, 4.01, 5.2
5) 2.1, 2.8, 3.8, 4.2, 5.7, 6.1
6) 0.14, 0.56, 0.87, 1.24, 3.1

Multiplying and Dividing Decimals

1) 3.15	6) 30.26	11) 3.336…	16) 0.041
2) 15.64	7) 43.68	12) 3.684…	17) 0.97
3) 15.75	8) 447	13) 2.705…	18) 5.24
4) 6.435	9) 54.9	14) 8.4	19) 1.65
5) 62.92	10) 4.117…	15) 0.0215	

Rounding Decimals

1) 0.3	7) 11.3	13) 6.5	19) 52
2) 5.0	8) 6	14) 1.6	20) 77
3) 8.8	9) 9.64	15) 73	21) 3
4) 0.5	10) 5.5	16) 84	22) 72
5) 7	11) 8	17) 41.7	23) 486.5
6) 0.3	12) 5	18) 795	24) 7.1

Comparing Decimals

1) 0.35 < 1.8	9) 7 > 0.7	17) 1.57 > 0.157
2) 1.9 > 1.19	10) 7.12 > 0.712	18) 0.092 > 0.091
3) 8.6 = 8.6	11) 12.3 < 12.5	19) 24.3 = 24.3
4) 2.45 < 24.5	12) 4.67 < 4.68	20) 0.17 < 0.71
5) 7.56 > 0.756	13) 2.57 < 2.75	21) 0.46 < 0.64
6) 11.4 > 11.05	14) 3.46 > 0.346	22) 0.2 > 0.08
7) 7.4 > 0.74	15) 6.87 > 6.78	23) 0.10 = 0.1
8) 8.56 > 0.85	16) 0.89 < 0.98	24) 3.52 < 31.5

Common Core Subject Test Mathematics Grade 5

Chapter 6 : Ratios and Rates

Topics that you'll learn in this chapter:

- ✓ Simplifying Ratios,
- ✓ Writing Ratios,
- ✓ Create a Proportion,
- ✓ Proportional Ratios,
- ✓ Similar Figures,
- ✓ Word Problems,

Common Core Subject Test Mathematics Grade 5

Simplifying Ratios

✏ Reduce each ratio.

1) 14: 56	9) 25: 20	17) 8: 88
2) 12: 36	10) 14: 28	18) 13: 52
3) 5: 35	11) 60: 84	19) 3: 75
4) 56: 48	12) 11: 99	20) 2.2: 22
5) 12: 14	13) 30: 45	21) 11: 33
6) 81: 63	14) 18: 45	22) 18: 81
7) 60: 3	15) 90: 15	23) 68: 80
8) 15: 10	16) 1.5: 3	24) 50: 500

Writing Ratios

✏ Express each ratio as a rate and unite rate.

1) 180 miles on 6 gallons of gas.

2) 99 dollars for 11 books.

3) 35 miles on 3.5 gallons of gas

4) 7.5 inches of snow in 1.5 hours

✏ Express each ratio as a fraction in the simplest form.

5) 6 feet out of 60 feet	11) 35 miles out of 120 miles
6) 14 cakes out of 49 cakes	12) 22 blue cars out of 55 cars
7) 32 dimes t0 60 dimes	13) 6.9 pennies to 69 pennies
8) 18 dimes out of 63 coins	14) 14 beetles out of 70 insects
9) 13 cups to 91 cups	15) 18 dimes to 54 dimes
10) 28 gallons to 42 gallons	16) 40 red cars out of 160 cars

WWW.MathNotion.Com

Common Core Subject Test Mathematics Grade 5

Create a Proportion

✏️ Create proportion from the given set of numbers.

1) 1, 20, 4, 5

2) 9, 135, 1, 15

3) 5, 15, 8, 24

4) 49, 7, 4, 28

5) 9, 1, 108, 12

6) 45, 2, 5, 18

7) 28, 7, 24, 6

8) 11, 3, 55, 15

9) 5, 45, 36, 4

10) 16, 128, 1, 8

11) 2.5, 10, 5, 20

12) 5, 12, 15, 36

Proportional Ratios

✏️ Solve each proportion.

1) $\dfrac{4}{8} = \dfrac{5}{d}$

2) $\dfrac{k}{6} = \dfrac{3}{9}$

3) $\dfrac{10}{8} = \dfrac{12}{x}$

4) $\dfrac{x}{15} = \dfrac{9}{5}$

5) $\dfrac{d}{11} = \dfrac{10}{1.1}$

6) $\dfrac{4.5}{6} = \dfrac{9}{x}$

7) $\dfrac{7}{15} = \dfrac{k}{60}$

8) $\dfrac{11}{1.5} = \dfrac{121}{d}$

9) $\dfrac{x}{0.7} = \dfrac{15}{2.8}$

10) $\dfrac{1.2}{4} = \dfrac{x}{2.5}$

11) $\dfrac{7.8}{x} = \dfrac{7.8}{2}$

12) $\dfrac{x}{3.4} = \dfrac{48}{16}$

13) $\dfrac{80}{20} = \dfrac{k}{60}$

14) $\dfrac{1.4}{5} = \dfrac{28}{d}$

15) $\dfrac{x}{7} = \dfrac{30}{15}$

16) $\dfrac{4}{1.6} = \dfrac{k}{1.6}$

17) $\dfrac{0.8}{1.2} = \dfrac{5.6}{d}$

18) $\dfrac{25}{x} = \dfrac{50}{4}$

19) $\dfrac{d}{9} = \dfrac{18}{27}$

20) $\dfrac{k}{12.6} = \dfrac{5}{12.6}$

21) $\dfrac{1.4}{7} = \dfrac{x}{10}$

22) $\dfrac{13}{5} = \dfrac{k}{15}$

23) $\dfrac{18}{21} = \dfrac{x}{7}$

24) $\dfrac{9}{99} = \dfrac{x}{22}$

Common Core Subject Test Mathematics Grade 5

Similar Figures

Each pair of figures is similar. Find the missing side.

1)

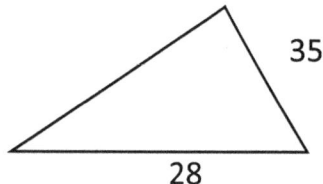

2)

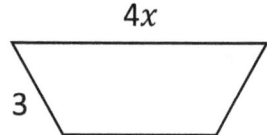 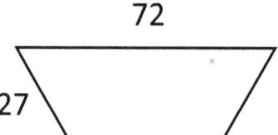

3)

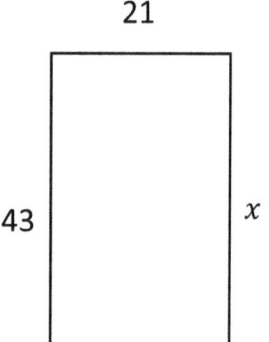 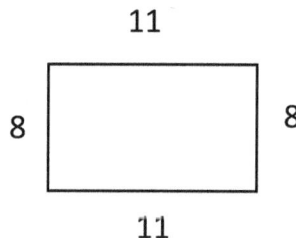

Word Problems

✎ Solve.

1) In a party, 15 soft drinks are required for every 18 guests. If there are 360 guests, how many soft drinks is required?

2) In Jack's class, 16 of the students are tall and 10 are short. In Michael's class 40 students are tall and 25 students are short. Which class has a higher ratio of tall to short students?

3) Are these ratios equivalent?

12 cards to 84 animals 18 marbles to 126 marbles

4) The price of 6 apples at the Quick Market is $2.7. The price of 7 of the same apples at Walmart is $3.64. Which place is the better buy?

5) The bakers at a Bakery can make 160 bagels in 4 hours. How many bagels can they bake in 11 hours? What is that rate per hour?

Common Core Subject Test Mathematics Grade 5

✎ Answer each question and round your answer to the nearest whole number.

6) If a 24.6 ft tall flagpole casts a 190.95 ft long shadow, then how long is the shadow that a 4.7 ft tall woman casts?

7) A model igloo has a scale of 2 in: 7 ft. If the real igloo is 56 ft wide, then how wide is the model igloo?

8) If an 88 ft tall tree casts a 8 ft long shadow, then how tall is an adult giraffe that casts a 4 ft shadow?

9) Find the distance between San Joe and Mount Pleasant if they are 5 cm apart on a map with a scale of 1 cm: 8 km.

10) A telephone booth that is 54 ft tall casts a shadow that is 9 ft long. Find the height of a lawn ornament that casts a 7 ft shadow.

Answers of Worksheets

Simplifying Ratios

1) 2: 8
2) 1: 3
3) 1: 7
4) 7: 6
5) 6: 7
6) 9: 7
7) 20: 1
8) 3: 2
9) 5: 4
10) 1: 2
11) 5: 7
12) 1: 9
13) 2: 3
14) 2: 5
15) 6: 1
16) 1: 2
17) 1: 11
18) 1: 4
19) 1: 25
20) 1: 10
21) 1: 3
22) 2: 9
23) 17: 20
24) 1: 10

Writing Ratios

1) $\frac{180 \text{ miles}}{6 \text{ gallons}}$, 30 miles per gallon
2) $\frac{99 \text{ dollars}}{11 \text{ books}}$, 9.00 dollars per book
3) $\frac{35 \text{ miles}}{3.5 \text{ gallons}}$, 10 miles per gallon
4) $\frac{7.5" \text{ of snow}}{1.5 \text{ hours}}$, 5 inches of snow per hour
5) $\frac{1}{10}$
6) $\frac{2}{7}$
7) $\frac{8}{15}$
8) $\frac{2}{7}$
9) $\frac{1}{7}$
10) $\frac{2}{3}$
11) $\frac{7}{24}$
12) $\frac{2}{5}$
13) $\frac{1}{10}$
14) $\frac{1}{5}$
15) $\frac{1}{3}$
16) $\frac{1}{4}$

Create Proportion

1) 1: 5 = 4: 20
2) 9: 135 = 1: 15
3) 8: 5 = 24: 15
4) 7: 4 = 49: 28
5) 9: 1 = 108: 12
6) 45: 18 = 5: 2
7) 24: 28 = 6: 7
8) 11: 3 = 55: 15
9) 4: 5 = 36: 45
10) 128: 16 = 8: 1
11) 2.5: 5 = 10: 20
12) 15: 5 = 36: 12

Proportional Ratios

1) 10
2) 2
3) 9.6
4) 27
5) 100
6) 12
7) 28
8) 16.5
9) 3.75
10) 0.75
11) 2
12) 10.2
13) 240
14) 100
15) 14
16) 4
17) 8.4
18) 2
19) 6
20) 5
21) 2
22) 39
23) 6
24) 2

Similar Figures

1) 4
2) 2
3) 43

Common Core Subject Test Mathematics Grade 5

Word Problems

1) 300

2) The ratio for both classes is equal to 8 to 5.

3) Yes! Both ratios are 1 to 7

4) The price at the Quick Market is a better buy.

5) 440, the rate is 40 per hour.

6) 36.48 ft

7) 16 in

8) 44 ft

9) 40 km

10) 42 ft

Common Core Subject Test Mathematics Grade 5

Chapter 7 : Measurement

Topics that you'll learn in this chapter:

- ✓ Reference Measurement Units,
- ✓ Metric Length Units,
- ✓ Customary Length Units,
- ✓ Metric Capacity Units,
- ✓ Customary Capacity Units,
- ✓ Metric Weight and Mass Units,
- ✓ Customary Weight and Mass Units,
- ✓ Temperature Units,
- ✓ Time,
- ✓ Add Money Amounts,
- ✓ Subtract Money Amounts,
- ✓ Money: Word Problems,

Common Core Subject Test Mathematics Grade 5

Reference Measurement Units

LENGTH

Customary

1 mile (mi) = 1,760 yards (yd)

1 yard (yd) = 3 feet (ft)

1 foot (ft) = 12 inches (in.)

Metric

1 kilometer (km) = 1,000 meters (m)

1 meter (m) = 100 centimeters (cm)

1 centimeter (cm) = 10 millimeters (mm)

VOLUME AND CAPACITY

Customary

1 gallon (gal) = 4 quarts (qt)

1 quart (qt) = 2 pints (pt.)

1 pint (pt.) = 2 cups (c)

1 cup (c) = 8 fluid ounces (Fl oz)

Metric

1 liter (L) = 1,000 milliliters (mL)

WEIGHT AND MASS

Customary

1 ton (T) = 2,000 pounds (lb.)

1 pound (lb.) = 16 ounces (oz)

Metric

1 kilogram (kg) = 1,000 grams (g)

1 gram (g) = 1,000 milligrams (mg)

Time

1 year = 12 months

1 year = 52 weeks

1 week = 7 days

1 day = 24 hours

1 hour = 60 minutes

1 minute = 60 seconds

WWW.MathNotion.Com

Common Core Subject Test Mathematics Grade 5

Metric Length Units

✏️ Convert to the units.

1) 300 mm = _____ cm

2) 8 m = _____ mm

3) 4.5 m = _____ cm

4) 7 km = _____ m

5) 9,400 mm = _____ m

6) 1,100 cm = _____ m

7) 2.8 m = _____ cm

8) 4,000 mm = _____ cm

9) 7,000 mm = _____ m

10) 2 km = _____ mm

11) 14.9 km = _____ m

12) 20 m = _____ cm

13) 5,000 m = _____ km

14) 7,600 m = _____ km

Customary Length Units

✏️ Convert to the units.

1) 8 ft = _____ in

2) 4 ft = _____ in

3) 6 yd = _____ ft

4) 10 yd = _____ ft

5) 3,520 yd = _____ mi

6) 60 in = _____ ft

7) 144 in = ____ yd

8) 0.5 mi = _____ yd

9) 15 yd = _____ in

10) 42 yd = _____ in

11) 99 ft = _____ yd

12) 1.5 mi = _____ yd

13) 84 in = _____ ft

14) 30 yd = _____ feet

Common Core Subject Test Mathematics Grade 5

Metric Capacity Units

✎ Convert the following measurements.

1) 32.4 l = _____ ml

2) 7.1 l = _____ ml

3) 54 l = _____ ml

4) 92 l = _____ ml

5) 48 l = _____ ml

6) 13 l = _____ ml

7) 750 ml = _____ l

8) 2,400 ml = _____ l

9) 73,000 ml = _____ l

10) 8,000 ml = _____ l

11) 49,000 ml = _____ l

12) 5,500 ml = _____ l

Customary Capacity Units

✎ Convert the following measurements.

1) 51 gal = _____ qt.

2) 35 gal = _____ pt.

3) 68 gal = _____ c.

4) 20 pt. = _____ c

5) 12.5 qt = _____ pt.

6) 22.5 qt = _____ c

7) 51 pt. = _____ c

8) 48 c = _____ gal

9) 96 pt. = _____ gal

10) 136 qt = _____ gal

11) 15 c = _____ fl oz

12) 44 c = _____ qt

13) 240 c = _____ pt.

14) 148 qt = _____ gal

15) 160 pt. = _____ qt

16) 104 fl oz = _____ c.

Common Core Subject Test Mathematics Grade 5

Metric Weight and Mass Units

✎ Convert.

1) 60 kg = _____ g

2) 24 kg = _____ g

3) 610 kg = _____ g

4) 82 kg = _____ g

5) 95.8 kg = _____ g

6) 4.85 kg = _____ g

7) 1.5 kg = _____ g

8) 95,000 g = _____ kg

9) 241,000 g = _____ kg

10) 700,000 g = _____ kg

11) 2,500 g = _____ kg

12) 28,900 g = _____ kg

13) 970,000 g = _____ kg

14) 325,500 g = _____ kg

Customary Weight and Mass Units

✎ Convert.

1) 10,000 lb. = _____ T

2) 19,000 lb. = _____ T

3) 32,000 lb. = _____ T

4) 16,800 lb. = _____ T

5) 27 lb. = _____ oz

6) 25.4 lb. = _____ oz

7) 124 lb. = _____ oz

8) 4T = _____ lb.

9) 7T = _____ lb.

10) 11.2T = _____ lb.

11) 12.8T = _____ lb.

12) $\frac{6}{5}$ T = _____ oz

13) 9.125 T = _____ oz

14) $\frac{3}{4}$ T = _____ oz

Common Core Subject Test Mathematics Grade 5

Temperature Units

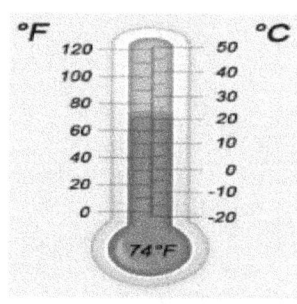

✏ Convert Fahrenheit into Celsius.

1) 21°F = ___ °C

2) 5.5°F = ___ °C

3) 71.6°F = ___ °C

4) 248°F = ___ °C

5) 75.2°F = ___ °C

6) 138.2°F = ___ °C

7) 167°F = ___ °C

8) 215.6°F = ___ °C

9) 104°F = ___ °C

10) 194°F = ___ °C

11) 203°F = ___ °C

12) 78.8°F = ___ °C

✏ Convert Celsius into Fahrenheit.

13) 2°C = ___ °F

14) 19°C = ___ °F

15) 48°C = ___ °F

16) 54°C = ___ °F

17) 81°C = ___ °F

18) 24°C = ___ °F

19) 88°C = ___ °F

20) 72°C = ___ °F

21) 80°C = ___ °F

22) 16°C = ___ °F

23) 10°C = ___ °F

24) 110°C = ___ °F

WWW.MathNotion.Com

Common Core Subject Test Mathematics Grade 5

Time

✏ Convert to the units.

1) 22 hr. = _____ min

2) 12.5 year = _____ week

3) 5.2 hr = _____ sec

4) 21 min = _____ sec

5) 1,800 min = _____ hr.

6) 730 day = _____ year

7) 1.5 year = _____ hr.

8) 42 day = _____ hr.

9) 2 day = _____ min

10) 660 min = _____ hr.

11) 10 year = _____ month

12) 2,124 sec = _____ min

13) 168 hr = _____ day

14) 19 weeks = _____ day

✏ How much time has passed?

1) From 2:25 A.M. to 5:35 A.M.: ____ hours and ____ minutes.

2) From 3:30 A.M. to 7:55 A.M.: ____ hours and ____ minutes.

3) It's 6:30 P.M. What time was 2 hours ago? _____ O'clock

4) 4:30 A.M to 7:50 AM: ____ hours and ____ minutes.

5) 4:15 A.M to 8:35 AM: ____ hours and ____ minutes.

6) 5:10 A.M. to 7:35 AM. = ____ hour(s) and ____ minutes.

7) 10:55 A.M. to 3:25 PM. = ____ hour(s) and ____ minutes

8) 8:05 A.M. to 8:40 A.M. = _____ minutes

9) 6:02 A.M. to 6:49 A.M. = _____ minutes

Money Amounts

Add.

1) $128 + $328 $225 + $245 $150 + $186

2) $453 + $128 $440 + $541 $258 + $248

3) $645 + $112.5 $235.4 + $452.1 $125.99 + $148.32

4) $321.40 + $175.80 $458.10 + $752.65 $652.00 + $324.70

Subtract.

5) $725 − $334 $543 − $248 $349 − $122

6) $658.20 − $220.30 $752.10 − $452.15 $312.50 − $89.90

7) $315.90 − $220.10 $548.40 − $342.10 $968.40 − $324.50

8) Linda had $18.60. She bought some game tickets for $9.25. How much did she have left?

Common Core Subject Test Mathematics Grade 5

Money: Word Problems

🖎 Solve.

1) How many boxes of envelopes can you buy with $40 if one box costs $8?

2) After paying $7.22 for a salad, Ella has $45.86. How much money did she have before buying the salad?

3) How many packages of diapers can you buy with $96 if one package costs $6?

4) Last week James ran 28.5 miles more than Michael. James ran 59 miles. How many miles did Michael run?

5) Last Friday Jacob had $14.68. Over the weekend he received some money for cleaning the attic. He now has $38.95. How much money did he receive?

6) After paying $3.15 for a sandwich, Amelia has $48.69. How much money did she have before buying the sandwich?

Common Core Subject Test Mathematics Grade 5

Answers of Worksheets

Metric length

1) 30 cm
2) 8,000 mm
3) 450 cm
4) 7,000 m
5) 9.4 m
6) 11 m
7) 280 cm
8) 400 cm
9) 7 m
10) 2,000,000 mm
11) 14,900 m
12) 2,000 cm
13) 5 km
14) 7.6 km

Customary Length

1) 96
2) 48
3) 18
4) 30
5) 2
6) 5
7) 4
8) 880
9) 540
10) 1,512
11) 33
12) 2,640
13) 7
14) 90

Metric Capacity

1) 32,400 ml
2) 7,100 ml
3) 54,000 ml
4) 92,000 ml
5) 48,000 ml
6) 13,000 ml
7) 0.75ml
8) 2.4 ml
9) 73 ml
10) 8L
11) 49 L
12) 5.5 L

Customary Capacity

1) 204 qt
2) 280 pt.
3) 1,088 c
4) 40 c
5) 25 pt.
6) 90c
7) 102 c
8) 3 gal
9) 12 gal
10) 34 gal
11) 120 qt
12) 11qt
13) 120 pt.
14) 37 gal
15) 80 qt
16) 13 pt.

Metric Weight and Mass

1) 60,000 g
2) 24,000 g
3) 610,000 g
4) 82,000 g
5) 95,800g
6) 4,850 g
7) 1,500 g
8) 95 kg
9) 241 kg
10) 700 kg
11) 2.5 kg
12) 28.9 kg
13) 970 kg
14) 325.5 kg

Customary Weight and Mass

1) 5 T
2) 9.5 T
3) 16 T
4) 8.4 T
5) 432 oz
6) 406.4 oz

WWW.MathNotion.Com

Common Core Subject Test Mathematics Grade 5

7) 1,984 oz
8) 8,000 lb.
9) 14,000 lb.
10) 22,400 lb.
11) 25,600 lb.
12) 38,400 oz
13) 292,000 oz
14) 24,000 oz

Temperature

1) −6.11°C
2) −14.72°C
3) 22°C
4) 120°C
5) 24°C
6) 59°C
7) 75°C
8) 102°C
9) 40°C
10) 90°C
11) 95°C
12) 26°C
13) 35.6°F
14) 66.2°F
15) 118.4°F
16) 129.2°F
17) 177.8°F
18) 75.2°F
19) 190.4°F
20) 161.6°F
21) 176°F
22) 60.8°F
23) 50°F
24) 230°F

Time - Convert

1) 1,320 min
2) 650 weeks
3) 18,720 sec
4) 1,260 sec
5) 30 hr
6) 2 year
7) 13,140 hr
8) 1,008 hr
9) 2,880 min
10) 11 hr
11) 120 months
12) 35.4 min
13) 7 days
14) 133 days

Time - Gap

1) 3:10
2) 4:25
3) 4:30 P.M.
4) 3:20
5) 4:20
6) 2:25
7) 4:30
8) 35 minutes
9) 47 minutes

Add Money

1) 456, 470, 336
2) 581, 981, 506
3) 757.5, 687.5, 274.31
4) 497.2, 1210.75, 976.7

Subtract Money

5) 391, 295, 227
6) 437.9, 299.95, 222.6
7) 95.8, 206.3, 643.9
8) $9.35

Money: word problem

1) 5
2) $53.08
3) 16
4) 30.5
5) 24.27
6) 51.84

Chapter 8 : Algebraic Thinking

Topics that you'll learn in this chapter:

- ✓ Finding Rules,
- ✓ Algebraic Word Problems,
- ✓ Evaluating Expressions,
- ✓ Variables and Expressions,

Common Core Subject Test Mathematics Grade 5

Finding Rules

✎ Complete the output.

1- **Rule:** the output is $x - 10.5$

Input	x	15	18	27	32.25	48.5
Output	y					

17) **Rule:** the output is $x \times 5\frac{1}{3}$

Input	x	3	9	15	21	33
Output	y					

2- **Rule:** the output is $x \div 9$

Input	x	513	387	342	198	126
Output	y					

✎ Find a rule to write an expression.

3- **Rule:** _____

Input	x	4	14	19	24
Output	y	10	35	47.5	60

4- **Rule:** _____

Input	x	5	13	19.6	34.5
Output	y	14.4	22.4	29	43.9

5- **Rule:** _____

Input	x	72	96	132	230.4
Output	y	9	12	16.5	28.8

Common Core Subject Test Mathematics Grade 5

Algebraic Word Problems

Circle the number sentence that fits the problem. Then solve for x.

1) Mary had $42. Then she earned more money (x). Now she has $86.

   $42 + x = $86   OR $42 + $86 = x

   x = ____

2) Lisa had $35. Then she earned more money (x). Now she has $78.

   $35 + x = $78   OR $35 + $78 = x

   x = ____

3) Matthew had $37. Then he earned more money (x). Now he has $98.

   $37 + x = $98   OR $37 + $98 = x

   x = ____

4) Charlotte gave 19 of the cookies he had baked to a friend and now he has 45 cookies left.   45 − 19 = x   OR   x − 19 = 45

   x = ____

5) Mia gave 32 of the cookies she had baked to a friend and now she has 55 cookies left.   55 − 32 = x   OR   x − 32 = 55

   x = ____

6) Lucas gave 41 of the cookies he had baked to a friend and now he has 49 cookies left.   . 49 − 41 = x   OR   x − 41 = 49

   x = ____

Common Core Subject Test Mathematics Grade 5

Evaluate Expressions

✏ Simplify each algebraic expression.

1) $11 - x$, $x = 3$

2) $x + 14$, $x = 4$

3) $8 - 3x$, $x = 1$

4) $3x + \frac{1}{3}$, $x = \frac{1}{2}$

5) $2x + 18$, $x = 1.5$

6) $7 - 3x$, $x = 1.2$

7) $12 + 2x - 15$, $x = 3.5$

8) $25 - 5x$, $x = 2.2$

9) $\frac{44}{x} - 58$, $x = 0.4$

10) $\frac{x}{3} - 15 + x$, $x = 12.6$

11) $\frac{x}{7} + 9.5$, $x = 28.7$

12) $\frac{33}{x} - 5.2 + 2.1x$, $x = 3$

13) $2x - \frac{45}{x} - 12$, $x = 9$

14) $\frac{x}{17} - 1.8$, $x = 34$

15) $2(12.5x + 8)$, $x = 2$

16) $16x + 13x - 27 + 9$,

$x = 1.5$

17) $8.7 - \frac{16}{x} + 3x$,

$x = 4$

18) $5(3a - 2a)$,

$a = 5.5$

19) $14 - 2x + 16 - x$,

$x = 3.5$

20) $6x - 3 - x$,

$x = 1.6$

21) $18 - 2(2x + x)$, $x = 0.2$

Variables and Expressions

✎ Write a verbal expression for each algebraic expression.

1) $3a - 7b$

2) $8.2c^2 + 5d$

3) $x - 19.5$

4) $\frac{90}{6}$

5) $x^2 + y^3$

6) $4x + 9$

7) $x^2 - 6y + 15$

8) $x^3 + 7y^2 - 6$

9) $\frac{1}{5}x + \frac{1}{4}y - 11$

10) $\frac{1}{7}(x + 13) - 18.6y$

✎ Write an algebraic expression for each verbal expression.

11) 14 less than h

12) The product of 19 and a

13) The 23.2 divided by K

14) The product of 7 and the third power of x

15) 14 more than h to the sixth power

16) 30 more than triple d

17) One eighth the square of b

18) The difference of 42.5 and 3 times a number

19) 73 more than the cube of a number

20) one-quarters the cube of a number

Common Core Subject Test Mathematics Grade 5

Answers of Worksheets

Finding Rules

1)

Input	x	15	18	27	32.25	48.5
Output	y	4.5	7.5	16.5	21.75	38

2)

Input	x	3	9	15	21	33
Output	y	16	48	80	112	176

3)

Input	x	513	387	342	198	126
Output	y	57	43	38	22	14

4) $y = 2.5x$ 5) $y = x + 9.4$ 6) $y = x \div 8$

Algebraic Word Problems

1) $\$42 + x = \$86$; $x = 44$ 4) $x - 19 = 45$; $x = 64$

2) $\$35 + x = \$78$; $x = 43$ 5) $x - 32 = 55$; $x = 87$

3) $\$37 + x = \$98$; $x = 61$ 6) $x - 41 = 49$; $x = 90$

Evaluating Expressions

1) 8   7) 4   13) 1   19) 19.5

2) 18   8) 14   14) 0.2   20) 5

3) 5   9) 52   15) 66   21) 16.8

4) $\frac{11}{6}$ 10) 1.8   16) 25.5

5) 21   11) 13.6   17) 16.7

6) 3.4   12) 12.1   18) 27.5

Variables and Expressions

1) 3 times a minus 7 times b.

2) 8.2 times c squared plus 5 times d.

3) a number minus 19.5.

4) the quotient of 90 and 6.

5) x squared plus y cubed.

6) the product of 4 and x plus 9.

Common Core Subject Test Mathematics Grade 5

7) x squared plus the product of 6 and y plus 15.

8) x cubed plus the product of 7 and y squared minus the product of 6 and y.

9) the sum of one–fifth of x and one–quarters of y, minus 11.

10) one–seventh of the sum of x and 13 minus the product of 18.6 and y.

11) $14 < h$

12) $19a$

13) $\frac{23.2}{K}$

14) $7x^3$

15) $14 > h^6$

16) $3d < 30$

17) $\frac{1}{8}b^2$

18) $42.5 - 3a$

19) $73 > a^3$

20) $\frac{1}{4}x^3$

Chapter 9 : Geometric

Topics that you'll learn in this chapter:

- ✓ Identifying Angles,
- ✓ Estimate and Measure Angles,
- ✓ Polygon Names,
- ✓ Classify Triangles,
- ✓ Parallel Sides in Quadrilaterals,
- ✓ Identify Parallelograms,
- ✓ Identify Trapezoids,
- ✓ Identify Rectangles,
- ✓ Perimeter and Area of Squares,
- ✓ Perimeter and Area of rectangles,
- ✓ Area and Perimeter: Word Problems,
- ✓ Circumference, Diameter and Radius,
- ✓ Volume of Cubes and Rectangle Prisms,

Identifying Angles

✏️ Write the name of the angles (Acute, Right, Obtuse, and Straight).

1)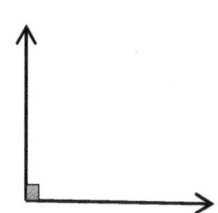

2)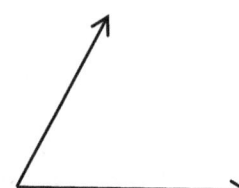

3)

4)

5)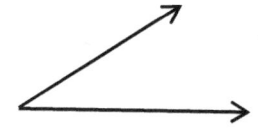

6)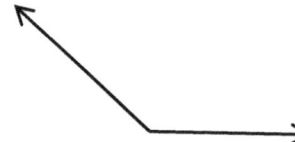

7)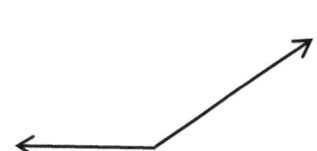

8)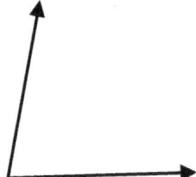

Common Core Subject Test Mathematics Grade 5

Estimate Angle Measurements

✏️ Estimate the approximate measurement of each angle in degrees.

1)

2)

3)

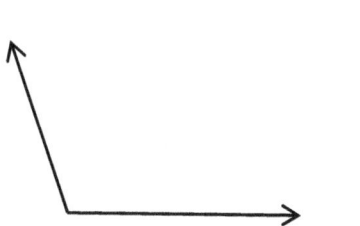

4)

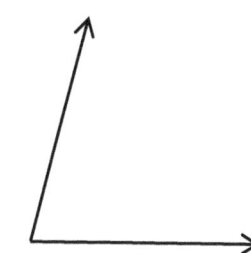

5)

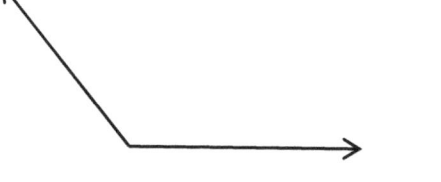

6)

7)

8)

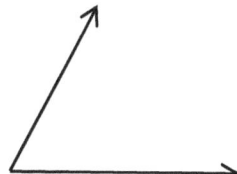

Common Core Subject Test Mathematics Grade 5

Measure Angles with a Protractor

✏️ Use protractor to measure the angles below.

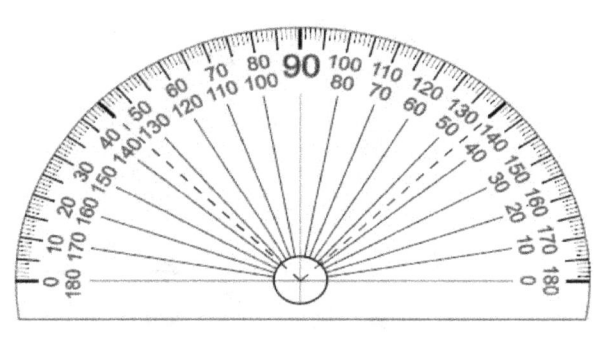

1) 2)

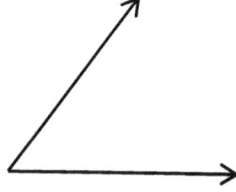   

3) 4)

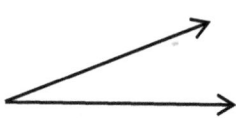   

✏️ Use a protractor to draw angles for each measurement given.

1) 140°

2) 100°

3) 110°

4) 120°

5) 55°

Polygon Names

✎ Write name of polygons.

1)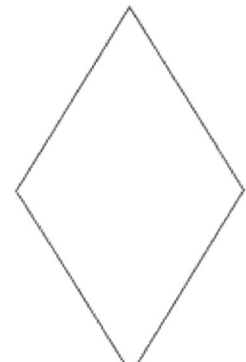

2)

3)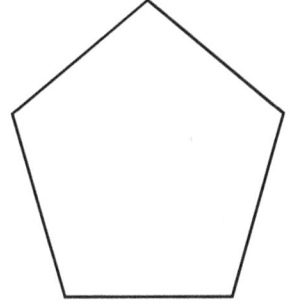

4)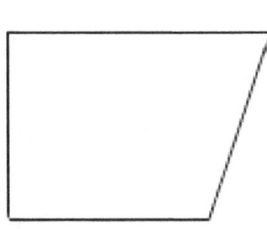

5)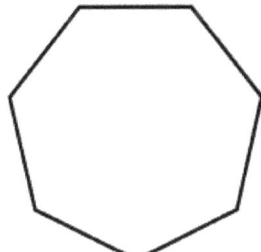

6)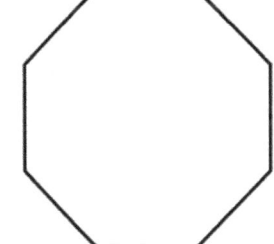

Classify Triangles

🖎 Classify the triangles by their sides and angles.

1)

2)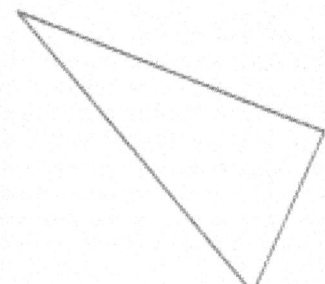

3)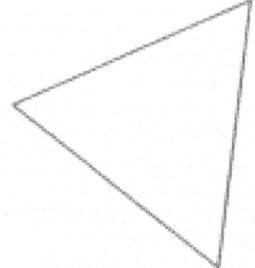

4)

5)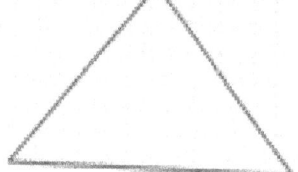

6)

Parallel Sides in Quadrilaterals

✏ Write name of quadrilaterals.

1)

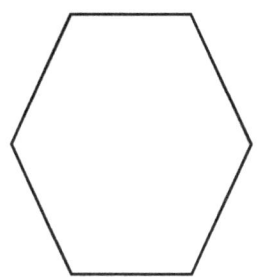

2)

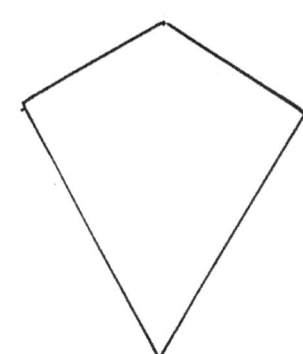

3)

4)

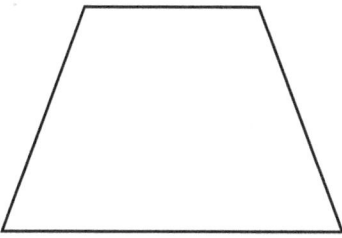

5)

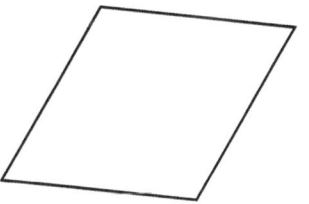

6)

Identify Rectangles

✏️ Solve.

1) A rectangle has _____ sides and _____ angles.

2) Draw a rectangle that is 5.5 centimeters long and 2.5 centimeters wide. What is the perimeter?

3) Draw a rectangle 3.5 cm long and 1.5 cm wide.

4) Draw a rectangle whose length is 4.25 cm and whose width is 2.45 cm. What is the perimeter of the rectangle?

5) What is the perimeter of the rectangle?

7.2

5.8

Common Core Subject Test Mathematics Grade 5

Perimeter: Find the Missing Side Lengths

✎ Find the missing side of each shape.

1) perimeter = 57.2

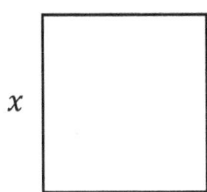

2) perimeter = 21.2

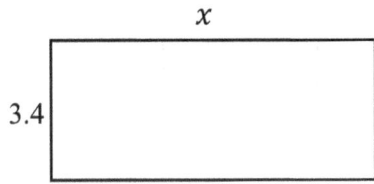

3) perimeter = 27.5

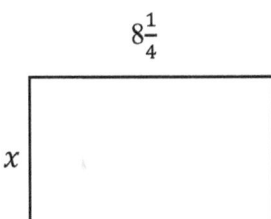

4) perimeter = 35.2

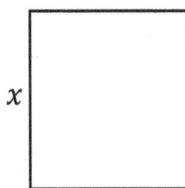

5) perimeter = 75.6

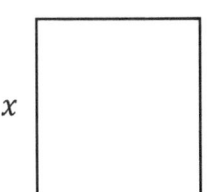

6) perimeter = 30.8

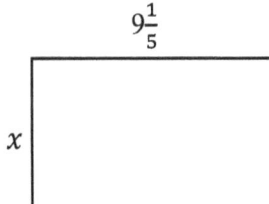

7) perimeter = 36.25

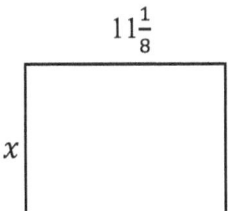

8) perimeter = 46.8

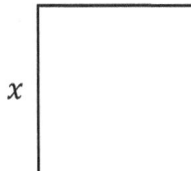

Common Core Subject Test Mathematics Grade 5

Perimeter and Area of Squares

✏️ Find perimeter and area of squares.

1) A: _____, P: _____

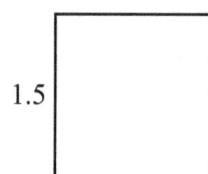

2) A: _____, P: _____

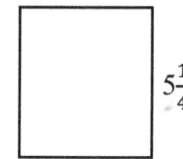

3) A: _____, P: _____

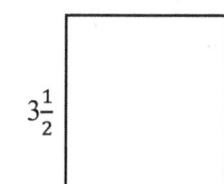

4) A: _____, P: _____

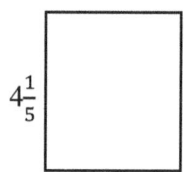

5) A: _____, P: _____

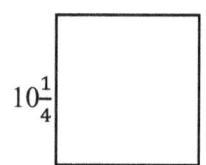

6) A: _____, P: _____

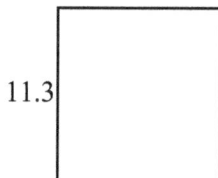

7) A: _____, P: _____

8) A: _____, P:

Common Core Subject Test Mathematics Grade 5

Perimeter and Area of rectangles

✎ Find perimeter and area of rectangles.

1) A: ____, P: ____

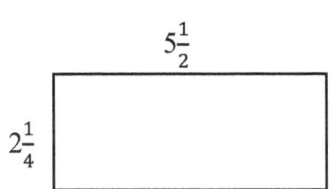

2) A: ____, P: ____

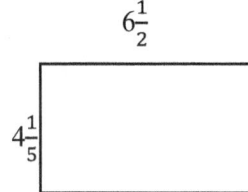

3) A: ____, P: ____

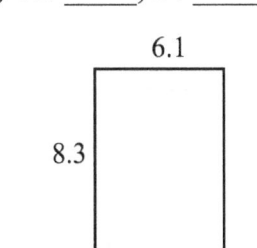

4) A: ____, P: ____

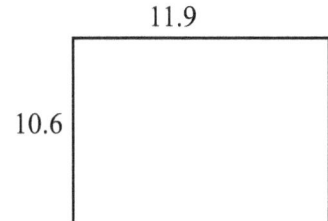

5) A: ____, P: ____

6) A: ____, P: ____

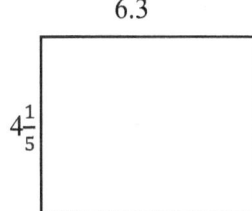

7) A: ____, P: ____

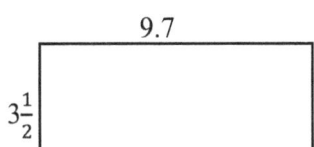

8) A: ____, P: ____

Common Core Subject Test Mathematics Grade 5

Find the Area or Missing Side Length of a Rectangle

✎ Find area or missing side length of rectangles.

1) Area =?

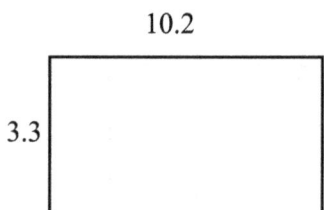

2) Area = 42.12, $x=$?

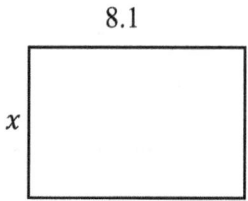

3) Area = 29.52, $x=$?

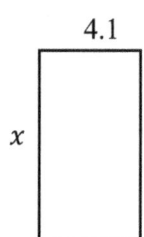

4) Area =?

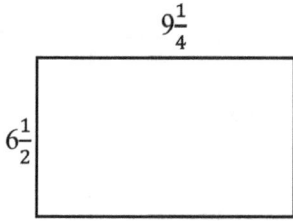

5) Area =?

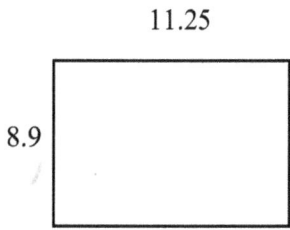

6) Area = 662.34 $x=$?

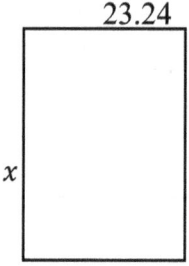

7) Area = 216.24, $x=$?

8) Area 336.42, $x=$?

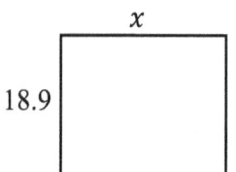

WWW.MathNotion.Com

Common Core Subject Test Mathematics Grade 5

Area and Perimeter: Word Problems

🖎 Solve.

1) The area of a rectangle is 90.86 square meters. The width is 7.7 meters. What is the length of the rectangle?

2) A square has an area of 6.25 square feet. What is the perimeter of the square?

3) Ava built a rectangular vegetable garden that is 3.2 feet long and has an area of 21.12 square feet. What is the perimeter of Ava's vegetable garden?

4) A square has a perimeter of 12.8 millimeters. What is the area of the square?

5) The perimeter of David's square backyard is 0.96 meters. What is the area of David's backyard?

6) The area of a rectangle is 37.63 square inches. The length is 7.1 inches. What is the perimeter of the rectangle?

Common Core Subject Test Mathematics Grade 5

Circumference, Diameter, and Radius

✏ Find the diameter and circumference of circles.

1)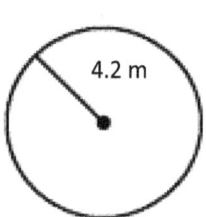

2)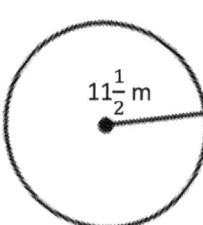

3)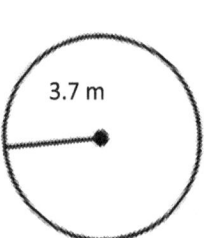

4)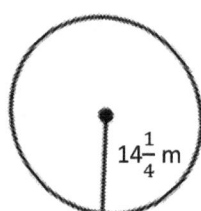

✏ Find the radius.

5)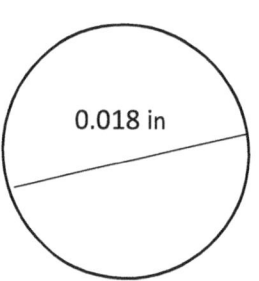

6)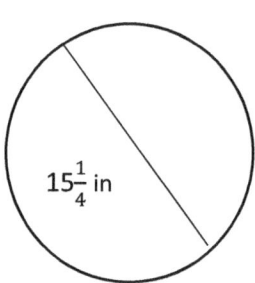

7) Diameter = $16\frac{1}{5} ft$

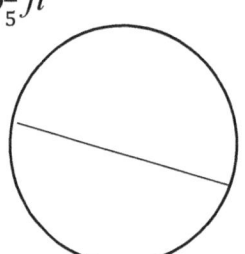

8) Diameter = 23.82 m

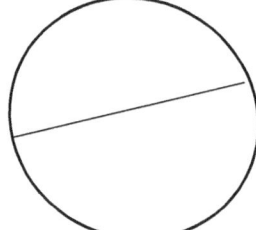

WWW.MathNotion.Com

Common Core Subject Test Mathematics Grade 5

Volume of Cubes and Rectangle Prisms

✎ Find the volume of each of the rectangular prisms.

1)

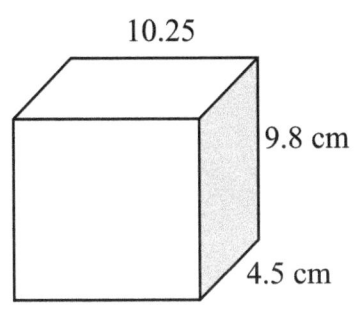

10.25
9.8 cm
4.5 cm

2)

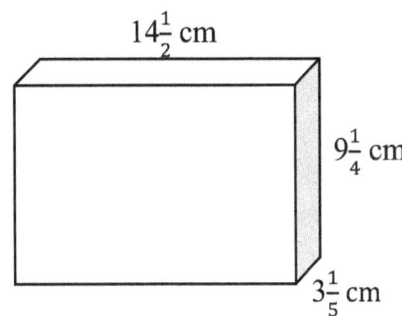

$14\frac{1}{2}$ cm
$9\frac{1}{4}$ cm
$3\frac{1}{5}$ cm

3)

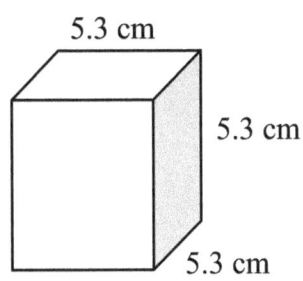

5.3 cm
5.3 cm
5.3 cm

4)

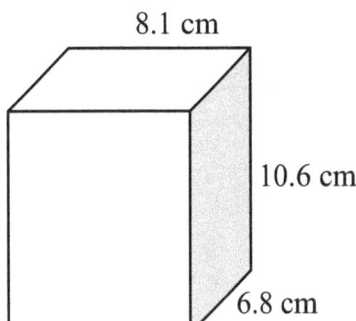

8.1 cm
10.6 cm
6.8 cm

5)

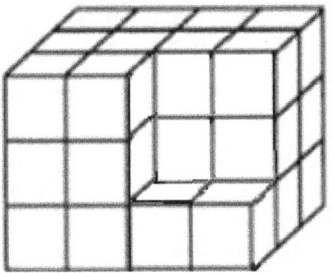

6)

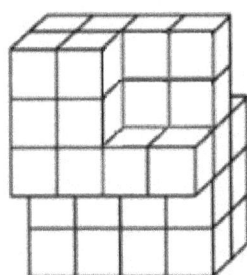

WWW.MathNotion.Com   103

Common Core Subject Test Mathematics Grade 5

Answers of Worksheets

Identifying Angles

1) Right   3) Obtuse   5) Acute   7) Obtuse
2) Acute   4) Straight   6) Obtuse   8) Acute

Estimate Angle Measurements

1) 160°   3) 110°   5) 130°   7) 90°
2) 180°   4) 75°   6) 45°   8) 60°

Measure Angles with a Protractor

1) 50°   2) 135°   3) 20°   4) 170°

Draw Angles

1)   2)   3)

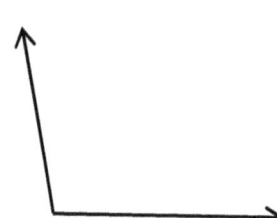 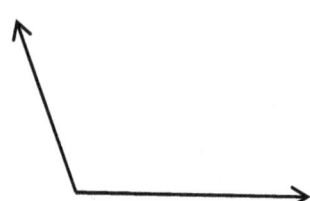

4)   5)

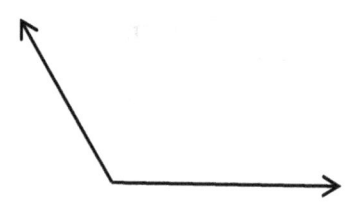 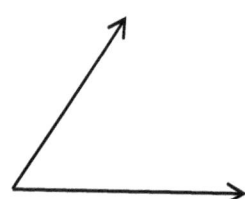

Polygon Names

1) Diamond   3) Pentagon   5) Heptagon
2) Parallelogram   4) Trapezius   6) Octagon

Classify Triangles

1) Scalene, acute   4) Scalene, right
2) Isosceles, acute   5) Isosceles, right
3) Equilateral, acute   6) Scalene, obtuse

WWW.MathNotion.Com

Common Core Subject Test Mathematics Grade 5

Parallel Sides in Quadrilaterals

1) Hexagon 3) Parallelogram 5) Rhombus
2) Kike 4) Trapezoid 6) Rectangle

Identify Rectangles

1) 4 - 4 3) Draw the rectangle. 5) 26
2) 16 4) 13.4

Perimeter: Find the Missing Side Lengths

1) 14.3 3) 5.5 5) 18.9 7) 7
2) 7.2 4) 8.8 6) 6.2 8) 11.7

Perimeter and Area of Squares

1) A: 2.25, P: 6 4) A: 17.64, P: 16.8 7) A: 153.76, P: 49.6
2) A: 27.56, P: 21 5) A: 105.063 P: 41 8) A: 96.04, P: 39.2
3) A: 12.25, P: 14 6) A: 127.69, P: 45.2

Perimeter and Area of rectangles

1) A: 12.375, P: 15.5 4) A: 126.14, P: 45 7) A: 33.95, P: 26.4
2) A: 27.3, P: 21.4 5) A: 39.9, P: 25.7 8) A: 66.12, P: 34.4
3) A: 50.63, P: 28.8 6) A: 26.46, P: 21

Find the Area or Missing Side Length of a Rectangle

1) 33.66 3) 7.2 5) 100.125 7) 10.2
2) 5.2 4) 60.125 6) 28.5 8) 17.8

Area and Perimeter: Word Problems

1) 11.8 3) 19.6 5) 0.0576
2) 10 4) 10.24 6) 24.8

Circumference, Diameter, and Radius

1) diameter: 8.4 circumferences: 8.4π or 26.376 4) diameter: 28.5 circumferences: 28.5π or
2) diameter: 23 circumferences: 23π or 72.22 89.49
3) diameter: 7.4 circumferences: 7.4π or 23.24

5) radius: 0.009 in 6) radius: 7.625 in 7) radius: 8.1 ft 8) radius: 11.91 m

Volume of Cubes and Rectangle Prisms

1) 452.025 cm³ 3) 148.877 c m³ 5) 32
2) 429.2 cm³ 4) 583.848 cm³ 6) 40

WWW.MathNotion.Com

Chapter 10 : Three-Dimensional Figures

Topics that you'll learn in this chapter:
- ✓ Identify Three–Dimensional Figures,
- ✓ Count Vertices, Edges, and Faces,
- ✓ Identify Faces of Three–Dimensional Figures,

Identify Three–Dimensional Figures

✎ Write the name of each shape.

1)

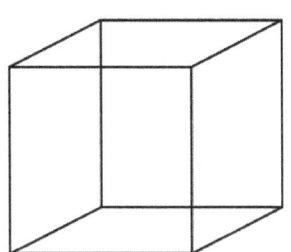

2)

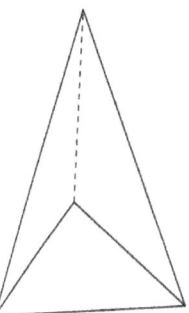

3)

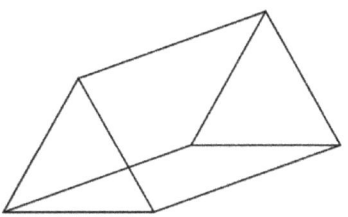

4)

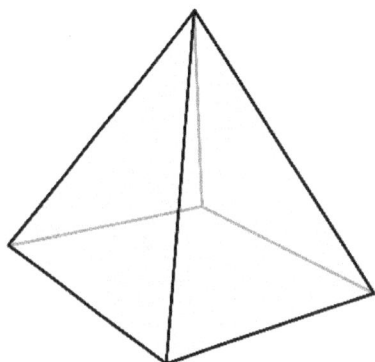

5)

6)

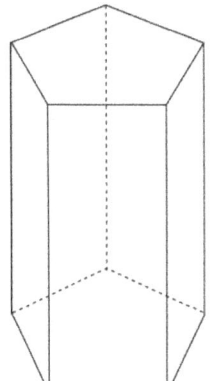

Common Core Subject Test Mathematics Grade 5

Count Vertices, Edges, and Faces

	Shape	Number of edges	Number of faces	Number of vertices
1)		___	___	___
2)		___	___	___
3)		___	___	___
4)		___	___	___
5)		___	___	___
6)		___	___	___

Common Core Subject Test Mathematics Grade 5

Identify Faces of Three-Dimensional Figures

✎ Write the number of faces.

1)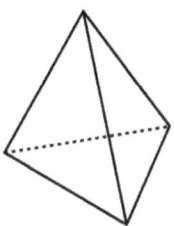

2)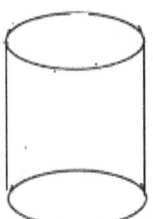

3)

4)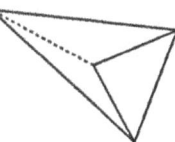

5)

6)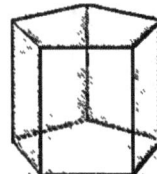

7)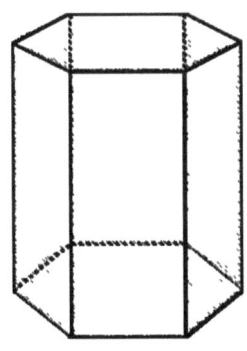

8)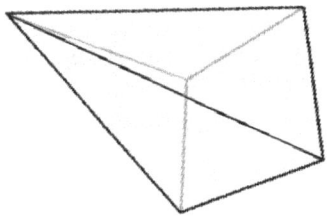

Common Core Subject Test Mathematics Grade 5

Answers of Worksheets

Identify Three–Dimensional Figures

1) Cube
2) Triangular pyramid
3) Triangular prism
4) Square pyramid
5) Rectangular prism
6) Pentagonal prism
7) Hexagonal prism

Count Vertices, Edges, and Faces

Shape	Number of edges	Number of faces	Number of vertices
1)	6	4	4
2)	8	5	5
3)	12	6	8
4)	15	7	10
5)	12	6	8
6)	18	8	12

Identify Faces of Three–Dimensional Figures

1) 6
2) 2
3) 5
4) 4
5) 6
6) 7
7) 8
8) 5

WWW.MathNotion.Com

Common Core Subject Test Mathematics Grade 5

Chapter 11 : Symmetry and Transformations

Topics that you'll learn in this chapter:
- ✓ Line Segments,
- ✓ Identify Lines of Symmetry,
- ✓ Count Lines of Symmetry,
- ✓ Parallel, Perpendicular and Intersecting Lines,

Common Core Subject Test Mathematics Grade 5

Line Segments

✎ Write each as a line, ray, or line segment.

1)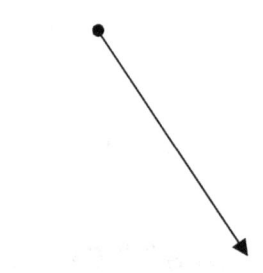

2)

3)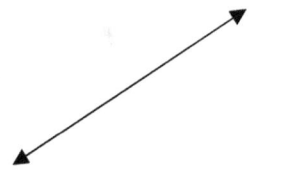

4)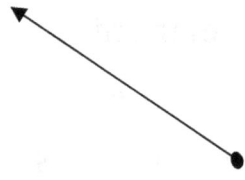

5)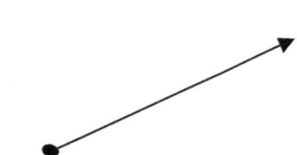

6)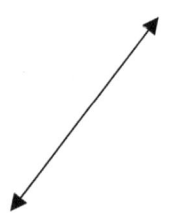

7)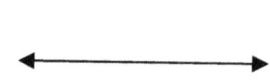

8)

WWW.MathNotion.Com

Common Core Subject Test Mathematics Grade 5

Identify Lines of Symmetry

✏️ Tell whether the line on each shape a line of symmetry is.

1)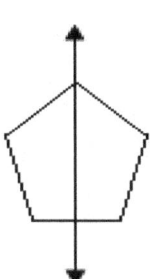

2)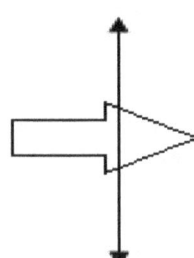

3)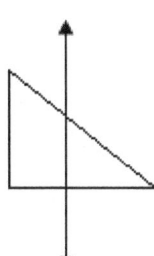

4)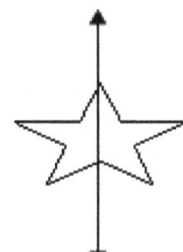

5)

6)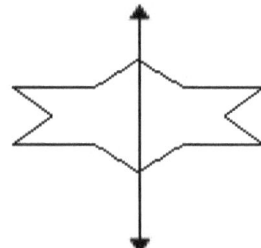

7)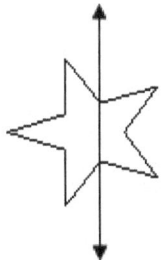

8)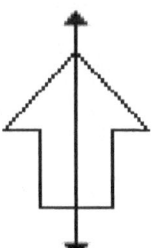

Count Lines of Symmetry

✎ Draw lines of symmetry on each shape. Count and write the lines of symmetry you see.

1)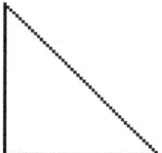

2)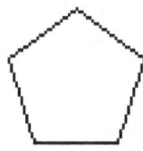

3)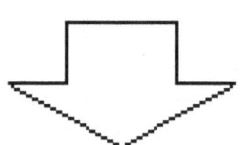

4)

5)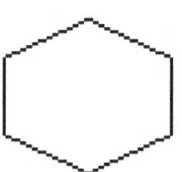

6)

7)

8)

Parallel, Perpendicular and Intersecting Lines

✍ State whether the given pair of lines are parallel, perpendicular, or intersecting.

1)

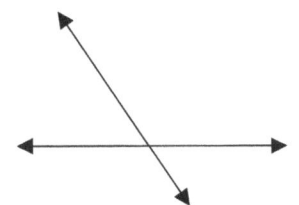

2)

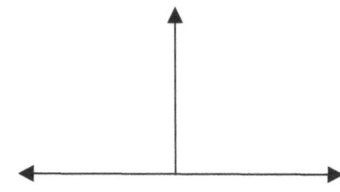

3)

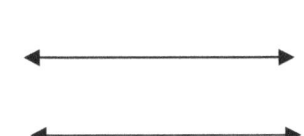

4)

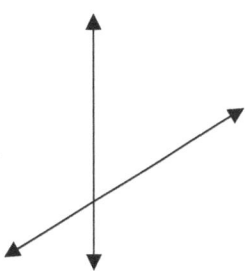

5)

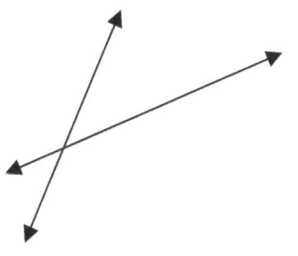

6)

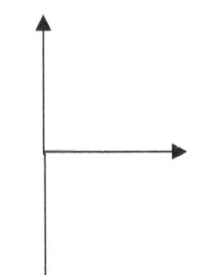

7)

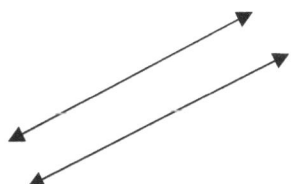

8)

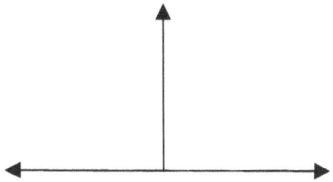

Common Core Subject Test Mathematics Grade 5

Answers of Worksheets

Line Segments

1) Ray
2) Line segment
3) Line
4) Ray
5) Ray
6) Line
7) Line
8) Line segment

Identify lines of symmetry

1) yes
2) no
3) no
4) yes
5) yes
6) yes
7) no
8) yes

Count lines of symmetry

1)

2)

3)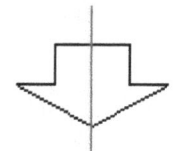

4)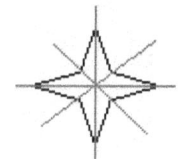

5)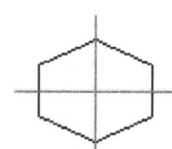

6)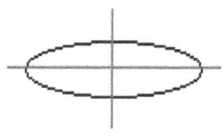

7)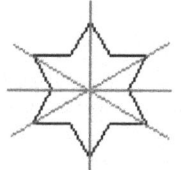

8)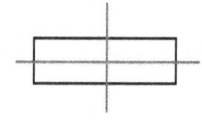

Parallel, Perpendicular and Intersecting Lines

1) Intersection
2) Perpendicular
3) Parallel
4) Intersection
5) Intersection
6) Perpendicular
7) Parallel
8) Perpendicular

Common Core Subject Test Mathematics Grade 5

Chapter 12 : Data Graphs, and Statistics

Topics that you'll learn in this chapter:

- ✓ Mean, Median, Mode, and Range,
- ✓ Graph Points on a Coordinate Plane,
- ✓ Bar Graph,
- ✓ Tally and Pictographs,
- ✓ Dot Plots
- ✓ Line Graphs,
- ✓ Stem–And–Leaf Plot,
- ✓ Scatter Plots,
- ✓ Probability Problems,

Common Core Subject Test Mathematics Grade 5

Mean and Median

🖉 Find Mean and Median of the Given Data.

1) 13, 22, 12, 6, 14

2) 7, 18, 11, 15, 2, 23

3) 33, 25, 14, 7, 14

4) 6, 7, 5, 1, 2, 5

5) 11, 4, 15, 7, 8, 17, 22

6) 9, 2, 5, 2, 18, 14, 45

7) 19, 14, 20, 12, 34, 14, 17, 31

8) 38, 29, 5, 3, 14, 9, 32

9) 28, 27, 31, 36, 32, 53, 41

10) 12, 17, 6, 17, 2, 17, 9, 12

11) 35, 11, 25, 54, 42, 41

12) 42, 48, 50, 48, 27, 67

13) 75, 72, 30, 46, 38, 29

14) 89, 73, 67, 46, 54, 84, 34

15) 78, 21, 100, 85, 54, 60

16) 41, 65, 9, 88, 17, 38, 14

🖉 Solve.

17) In a javelin throw competition, five athletics score 54, 72, 59, 67 and 85 meters. What are their Mean and Median? _____

18) Eva went to shop and bought 18 apples, 7 peaches, 6 bananas, 3 pineapple and 9 melons. What are the Mean and Median of her purchase?

_____

WWW.MathNotion.Com

Common Core Subject Test Mathematics Grade 5

Mode and Range

✎ Find Mode and Rage of the Given Data.

1) 12, 4, 8, 2, 9, 4
   Mode: _____ Range: _____

2) 8, 8, 11, 8, 18, 2, 5, 41
   Mode: _____ Range: _____

3) 3, 3, 2, 23, 4, 12, 3, 9, 3
   Mode: _____ Range: _____

4) 12, 39, 5, 25, 4, 4, 27, 8
   Mode: _____ Range: _____

5) 5, 5, 8, 5, 14, 3, 12
   Mode: _____ Range: _____

6) 2, 5, 19, 15, 12, 11, 7, 8, 7, 7
   Mode: _____ Range: _____

7) 9, 4, 0, 12, 19, 21, 9, 7, 9, 3
   Mode: _____ Range: _____

8) 11, 4, 3, 11, 5, 8, 44, 11, 7
   Mode: _____ Range: _____

9) 3, 3, 6, 9, 3, 3, 9, 15, 14, 20
   Mode: _____ Range: _____

10) 15, 14, 14, 16, 20, 7, 1, 30
    Mode: _____ Range: _____

11) 13, 3, 27, 4, 4, 16, 18, 4
    Mode: _____ Range: _____

12) 7, 22, 35, 11, 7, 24, 6, 13
    Mode: _____ Range: _____

✎ Solve.

13) A stationery sold 14 pencils, 28 red pens, 51 blue pens, 14 notebooks, 27 erasers, 43 rulers and 40 color pencils. What are the Mode and Range for the stationery sells?

    Mode: _____   Range: _____

14) In an English test, eight students score 14, 20, 22, 14, 17, 28, 36 and 24. What are their Mode and Range? _____

WWW.MathNotion.Com

Common Core Subject Test Mathematics Grade 5

Graph Points on a Coordinate Plane

✎ Plot each point on the coordinate grid.

1) A (4, 6) 3) C (1, 5) 5) E (4, 8)
2) B (3, 2) 4) D (5, 7) 6) F (9, 2)

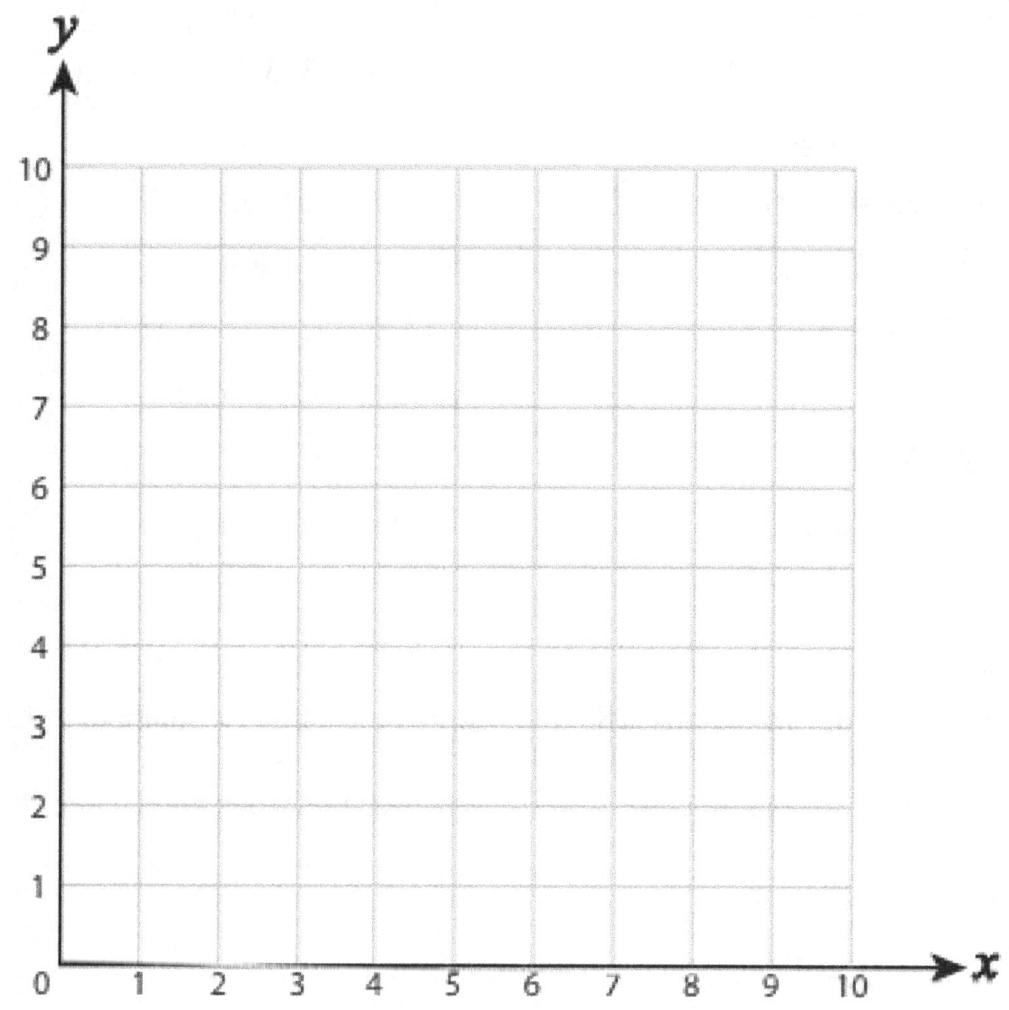

WWW.MathNotion.Com

Bar Graph

✏ Graph the given information as a bar graph.

Day	Hot dogs sold
Monday	50
Tuesday	80
Wednesday	10
Thursday	30
Friday	70

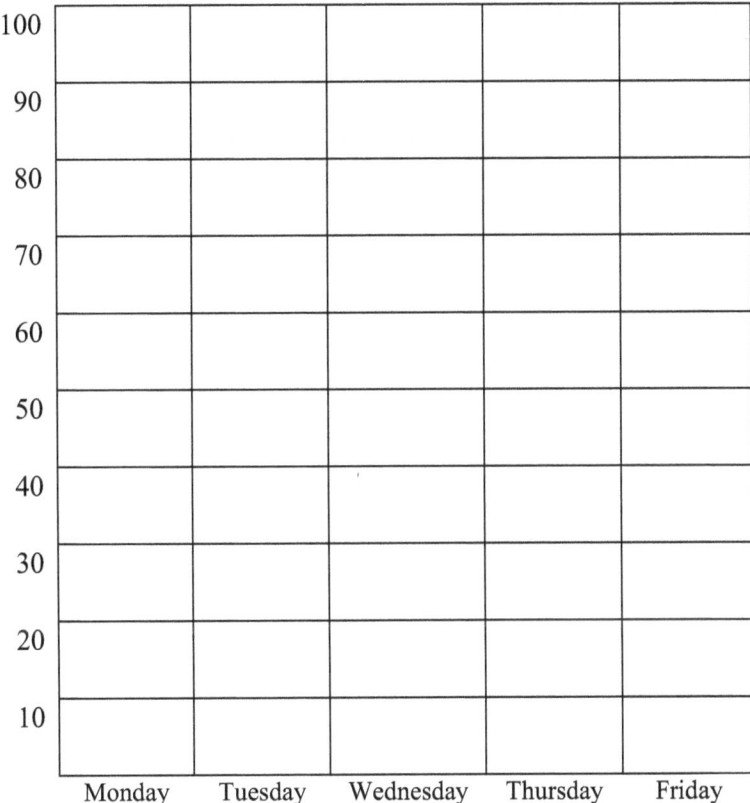

Tally and Pictographs

✎ Using the key, draw the pictograph to show the information.

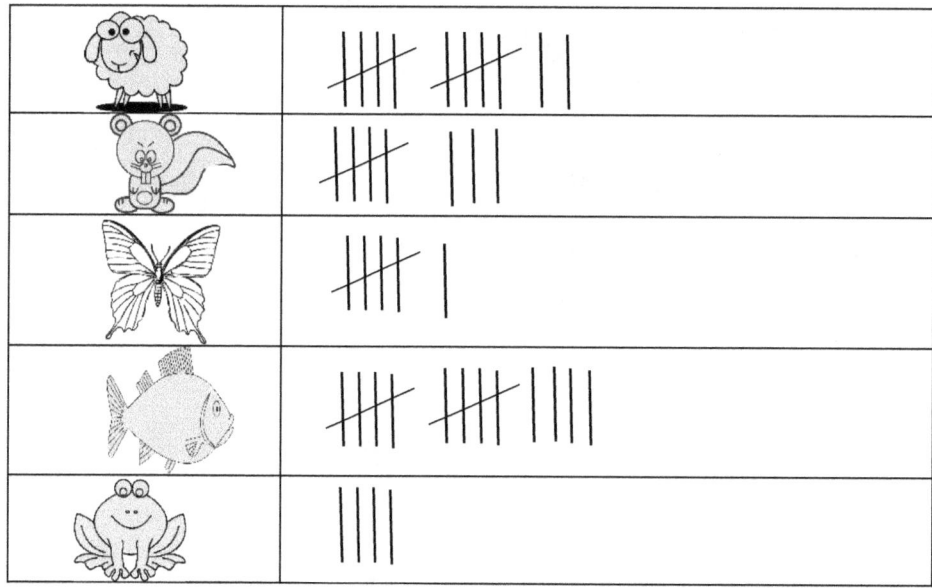

Key: ⚽ = 2 animals

Dot plots

The ages of students in a Math class are given below.

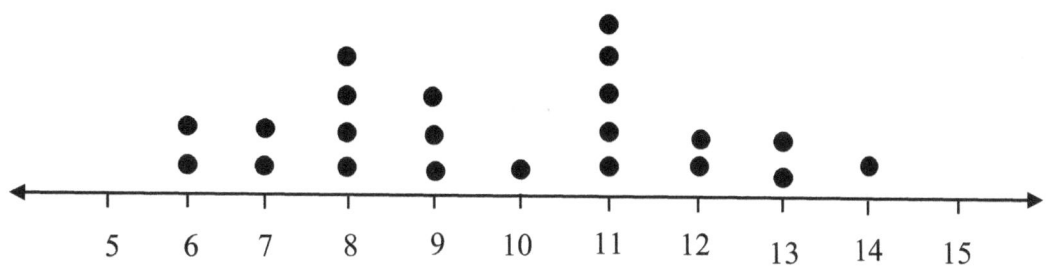

1) What is the total number of students in math class?

2) How many students are at least 12 years old?

3) Which age(s) has the most students?

4) Which age(s) has the fewest student?

5) Determine the median of the data.

6) Determine the range of the data.

7) Determine the mode of the data.

Line Graphs

🖊 David work as a salesman in a store. He records the number of shoes sold in five days on a line graph. Use the graph to answer the question

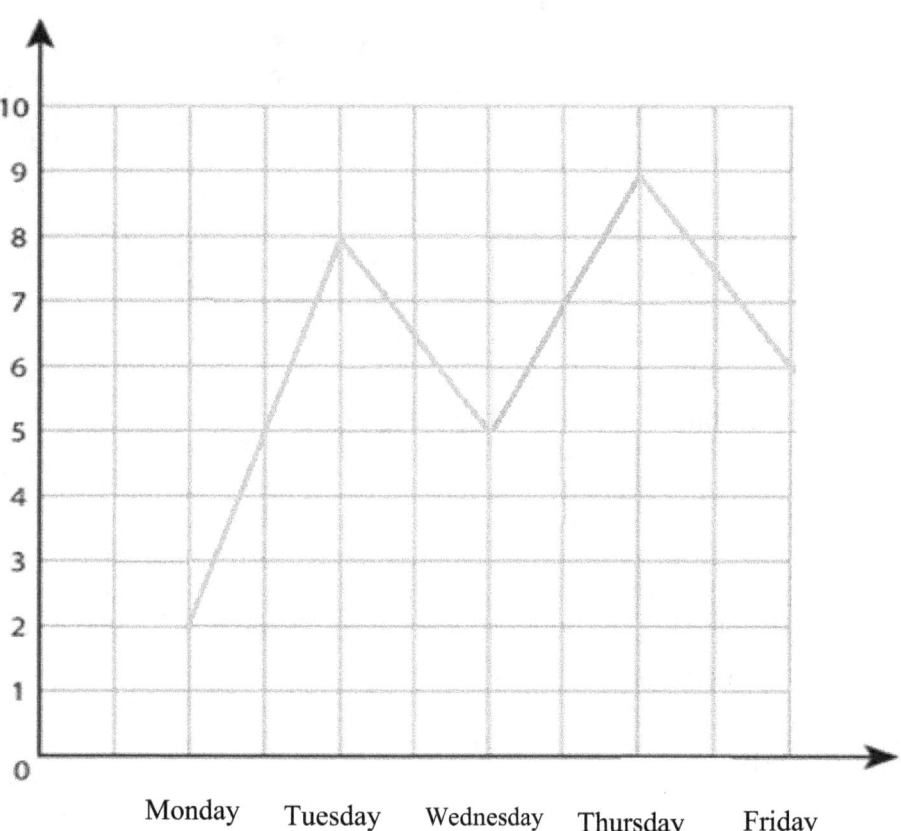

1) How many shoes were sold on Friday?

2) Which day had the minimum sales of shoes?

3) Which day had the maximum number of shoes sold?

4) How many shoes were sold in 5 days?

Stem-And-Leaf Plot

✎ Make stem ad leaf plots for the given data.

1) 42, 47, 14, 19, 42, 69, 65, 49, 42, 10, 64

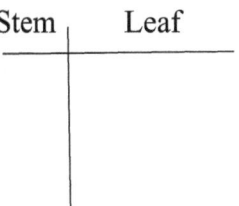

2) 43, 85, 52, 48, 45, 43, 51, 81, 59, 50, 85, 89

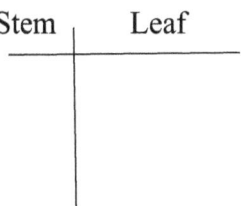

3) 112, 39, 46, 35, 80, 119, 42, 114, 37, 112, 47, 119

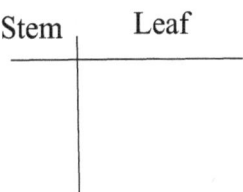

4) 90, 50, 131, 93, 112, 56, 139, 98, 115, 59, 98, 135, 111

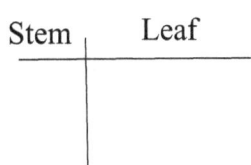

Common Core Subject Test Mathematics Grade 5

Scatter Plots

✏️ **Construct a scatter plot.**

x	1	2	3	4	5	8
y	15	40	50	35	70	20

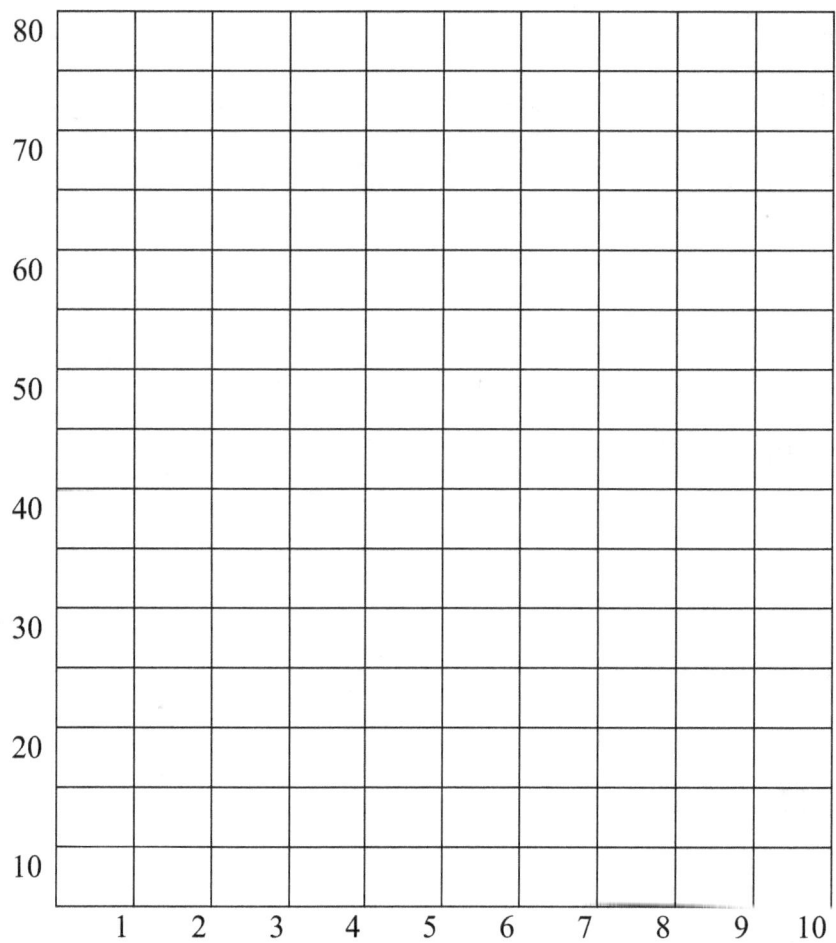

Probability Problems

✎ Solve.

1) A number is chosen at random from 1 to 20. Find the probability of selecting a 10 or smaller.

2) A number is chosen at random from 1 to 25. Find the probability of selecting multiples of 5.

3) A number is chosen at random from 1 to 15. Find the probability of selecting multiples of 7.

4) A number is chosen at random from 1 to 20. Find the probability of selecting a multiple of 6.

5) A number is chosen at random from 1 to 10. Find the probability of selecting prime numbers.

6) A number is chosen at random from 1 to 20. Find the probability of not selecting factors of 15.

Common Core Subject Test Mathematics Grade 5

Answers of Worksheets

Mean and Median

1) Mean: 13.4, Median: 13
2) Mean: 12.67, Median: 13
3) Mean: 18.6, Median: 14
4) Mean: 4.33, Median: 5
5) Mean: 12, Median: 11
6) Mean: 13.57, Median: 9
7) Mean: 20.125, Median: 18
8) Mean: 18.57, Median: 14
9) Mean: 35.43, Median: 32
10) Mean: 11.5, Median: 12
11) Mean: 34.67, Median: 38
12) Mean: 47, Median: 48
13) Mean: 48.33, Median: 42
14) Mean: 63.86, Median: 67
15) Mean: 66.33, Median: 69
16) Mean: 38.86, Median: 38
17) Mean: 67.4, Median: 67
18) Mean: 8.6, Median: 7

Mode and Range

1) Mode: 4, Range: 10
2) Mode: 8, Range: 39
3) Mode: 3, Range: 21
4) Mode: 4, Range: 35
5) Mode: 5, Range: 11
6) Mode: 7, Range: 17
7) Mode: 9, Range: 21
8) Mode: 11, Range: 41
9) Mode: 3, Range: 17
10) Mode: 14, Range: 29
11) Mode: 4, Range: 24
12) Mode: 7, Range: 29
13) Mode: 14, Range: 37
14) Mode: 14, Range: 22

Graph Points on a Coordinate Plane

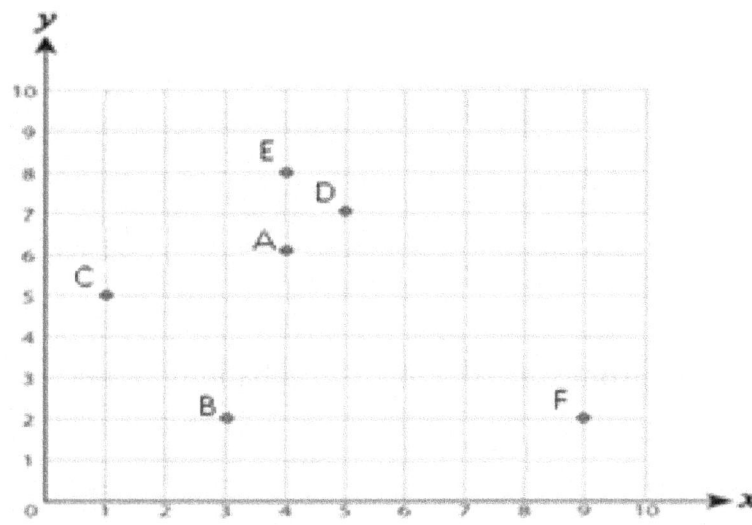

Common Core Subject Test Mathematics Grade 5

Bar Graph

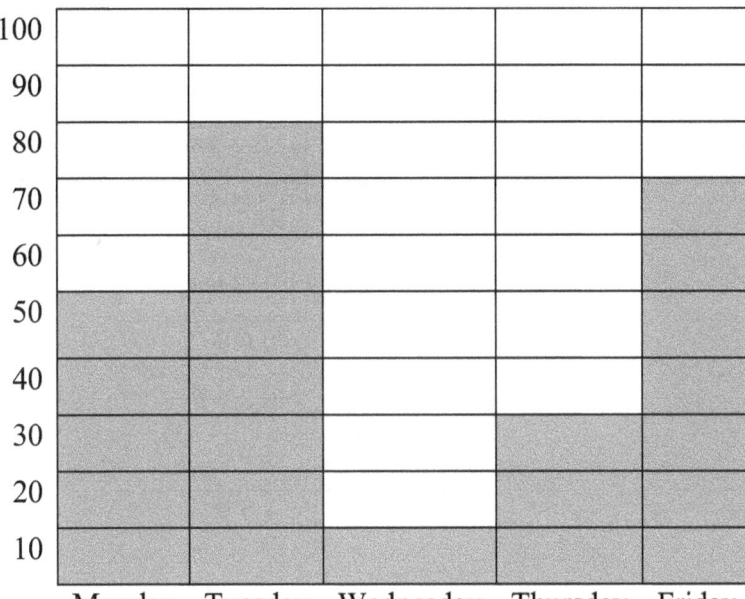

Tally and Pictographs

Dot plots

1) 22
2) 5
3) 11
4) 10 and 14
5) 2
6) 4
7) 2

Line Graphs

1) 6
2) Monday
3) Thursday
4) 30

Common Core Subject Test Mathematics Grade 5

Stem–And–Leaf Plot

1)
Stem	leaf
1	0 4 9
4	2 2 2 7 9
6	4 5 9

2)
Stem	leaf
4	3 3 5 8
5	0 1 2 9
8	1 5 5 9

3)
Stem	leaf
3	5 7 9
4	2 6 7
8	0
11	2 2 4 9 9

4)
Stem	leaf
5	0 6 9
9	0 3 8 8
11	1 2 5
13	1 5 9

Scatter Plots

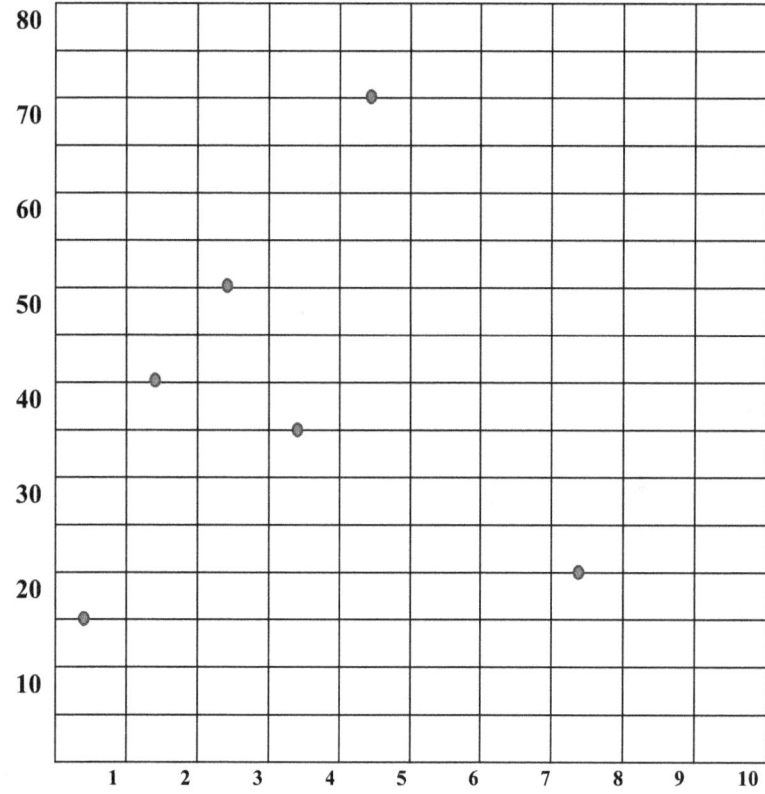

Probability Problems

1) $\frac{1}{2}$

2) $\frac{1}{5}$

3) $\frac{2}{15}$

4) $\frac{3}{20}$

5) $\frac{2}{5}$

6) $\frac{4}{5}$

Common Core Subject Test Mathematics Grade 5

Chapter 13 : Common Core Math Practice Tests

Time to Test

Time to refine your skill with a practice examination.

Take a REAL Common Core Mathematics test to simulate the test day experience. After you've finished, score your test using the answer key.

Before You Start

- You'll need a pencil and scratch papers to take the test.
- For this practice test, don't time yourself. Spend time as much as you need.
- It's okay to guess. You won't lose any points if you're wrong.
- After you've finished the test, review the answer key to see where you went wrong.

Calculators are not permitted for Grade 5 Common Core Tests

Good Luck!

Common Core Subject Test Mathematics Grade 5

Common Core GRADE 5 MAHEMATICS REFRENCE MATERIALS

Perimeter

Square $\qquad P = 4S$

Rectangle $\qquad P = 2L + 2W$

Area

Square $\qquad A = S \times S$

Rectangle $\qquad A = l \times w \qquad$ or $\qquad A = bh$

Volume

Square $\qquad A = S \times S \times S$

Rectangle $\qquad A = l \times w \times h \qquad$ or $\qquad A = Bh$

LENGTH

Customary	Metric
1 mile (mi) = 1,760 yards (yd)	1 kilometer (km) = 1,000 meters (m)
1 yard (yd) = 3 feet (ft)	1 meter (m) = 100 centimeters (cm)
1 foot (ft) = 12 inches (in.)	1 centimeter (cm) = 10 millimeters (mm)

VOLUME AND CAPACITY

Customary	Metric
1 gallon (gal) = 4 quarts (qt)	1 liter (L) = 1,000 milliliters (mL)
1 quart (qt) = 2 pints (pt.)	
1 pint (pt.) = 2 cups (c)	
1 cup (c) = 8 fluid ounces (Fl oz)	

WEIGHT AND MASS

Customary	Metric
1 ton (T) = 2,000 pounds (lb.)	1 kilogram (kg) = 1,000 grams (g)
1 pound (lb.) = 16 ounces (oz)	1 gram (g) = 1,000 milligrams (mg)

Common Core Practice Test 1

Mathematics

GRADE 5

Released Month Year

Common Core Subject Test Mathematics Grade 5

1) A rope 19 yards long is cut into 4 equal parts. Which expression does NOT equal to the length of each part?

   A. $19 \div 4$

   B. $\frac{19}{4}$

   C. $4 \div 19$

   D. $4\overline{)23}$

2) Calculate the area of the trapezoid in the following figure.

   A. 60 ft²

   B. 46 ft²

   C. 64 ft²

   D. 42 ft²

   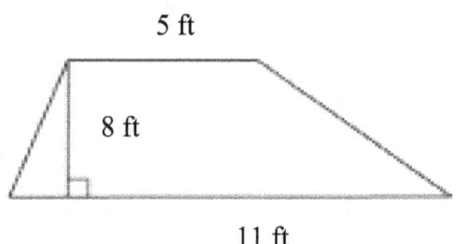

3) The drivers at G & G trucking must report the mileage on their trucks each week. The mileage reading of Ed's vehicle was 45,386 at the beginning of one week, and 45,613 at the end of the same week. What was the total number of miles driven by Ed that week?

   A. 347 Miles

   B. 229 Miles

   C. 227 Miles

   D. 327 Miles

Common Core Subject Test Mathematics Grade 5

4) Camille uses a 35% off coupon when buying a sweater that costs $60. How much does she pay?

   A. $39

   B. $18

   C. $28.55

   D. $32

5) A baker uses 7 eggs to bake a cake. How many cakes will he be able to bake with 910 eggs?

   A. 160

   B. 130

   C. 150

   D. 110

6) Which of the following angles is obtuse?

   A. 30 Degrees

   B. 45 Degrees

   C. 75 Degrees

   D. 115 Degrees

7) Which of the following fractions is the largest?

   A. $\frac{1}{4}$

   B. $\frac{3}{7}$

   C. $\frac{7}{8}$

   D. $\frac{4}{7}$

Common Core Subject Test Mathematics Grade 5

8) The area of a rectangle is D square feet and its length is 27 feet. Which equation represents W, the width of the rectangle in feet?

A. $W = \dfrac{D}{27}$

B. $W = \dfrac{27}{D}$

C. $W = 27D$

D. $W = 27 + D$

9) Which list shows the fractions in order from least to greatest?

$$\dfrac{2}{9}, \dfrac{1}{8}, \dfrac{7}{12}, \dfrac{8}{9}, \dfrac{5}{7}$$

A. $\dfrac{7}{12}, \dfrac{8}{9}, \dfrac{1}{8}, \dfrac{5}{7}, \dfrac{2}{9}$

B. $\dfrac{2}{9}, \dfrac{5}{7}, \dfrac{7}{12}, \dfrac{8}{9}, \dfrac{1}{8}$

C. $\dfrac{1}{8}, \dfrac{7}{12}, \dfrac{8}{9}, \dfrac{5}{7}, \dfrac{2}{9}$

D. $\dfrac{1}{8}, \dfrac{2}{9}, \dfrac{7}{12}, \dfrac{5}{7}, \dfrac{8}{9}$

10) Which statement about 6 multiplied by $\dfrac{8}{13}$ is true?

A. The product is between 2 and 3.

B. The product is between 3 and 4.

C. The product is more than $\dfrac{16}{7}$.

D. The product is between $\dfrac{9}{2}$ and 5.

11) What is the volume of this box?

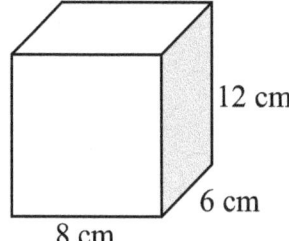

A. 543 cm³

B. 765 cm³

C. 620 cm³

D. 576 cm³

12) A shirt costing $360 is discounted 25%. Which of the following expressions can be used to find the selling price of the shirt?

A. (360) (0.15)

B. (360) – 360 (0.75)

C. (360) (0.75) – (360) (0.25)

D. (360) (0.75)

13) The perimeter of the trapezoid below is 44. What is its area?

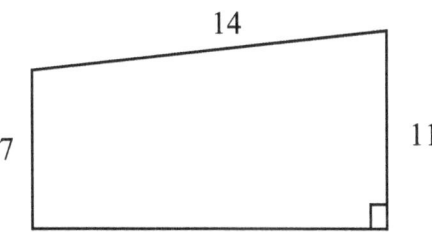

Write your answer in the box below.

Common Core Subject Test Mathematics Grade 5

14) In a bag, there are 105 cards. Of these cards, 15 cards are white. What fraction of the cards are white?

   A. $\frac{1}{7}$

   B. $\frac{7}{15}$

   C. $\frac{15}{22}$

   D. $\frac{1}{15}$

15) Which expression is equal to $\frac{9}{40}$?

   A. 40: 9

   B. 9 ÷ 40

   C. 9$\overline{)40}$

   D. 40 × 9

16) If A = 70, then which of the following equations are correct?

   A. A + 50 = 120

   B. A ÷ 50 = 120

   C. 50 × A = 120

   D. A − 50 = 120

17) 26 is What percent of 40?

   A. 32 %

   B. 90 %

   C. 65 %

   D. 75 %

Common Core Subject Test Mathematics Grade 5

18) How long does a 425–miles trip take moving at 50 miles per hour (mph)?

   A. 8 hours

   B. 9 hours and 10 minutes

   C. 8 hours and 50 minutes

   D. 8 hours and 30 minutes

19) The area of a circle is 25π. What is the circumference of the circle?

   A. 5π

   B. 10π

   C. 50π

   D. 20π

20) A rope weighs 420 grams per meter of length. What is the weight in kilograms of 50 meters of this rope? (1 kilograms = 1,000 grams)

   A. 0.021

   B. 0.21

   C. 21

   D. 21,000

21) How many $\frac{1}{8}$ cup servings are in a package of cheese that contains $4\frac{1}{4}$ cups altogether?

   Write your answer in the box below.

Common Core Subject Test Mathematics Grade 5

22) With what number must 8.6541654 be multiplied in order to obtain the number 86,541.654?

   A. 100

   B. 100,000

   C. 10,000

   D. 1,000,000

23) Lily and Ella are in a pancake–eating contest. Lily can eat eight pancakes per minute, while Ella can eat $3\frac{1}{6}$ pancakes per minute. How many total pancakes can they eat in 6 minutes?

   A. 58 Pancakes

   B. 19 Pancakes

   C. 48 Pancakes

   D. 67 Pancakes

24) The distance between cities A and B is approximately 3,198 miles. If Alice drives an average of 78 miles per hour, how many hours will it take Alice to drive from city A to city B?

   A. Approximately 51 Hours

   B. Approximately 41 Hours

   C. Approximately 31 Hours

   D. Approximately 21 Hours

Common Core Subject Test Mathematics Grade 5

25) 10 yards 8 feet and 18 inches equals to how many inches?

   A. 188

   B. 384

   C. 474

   D. 488

26) Which expression has a value of – 9?

   A. $21 - (-6) + (-36)$

   B. $4 + (-3) \times (-7)$

   C. $-8 \times (-2) - (-9) \times (-5)$

   D. $(-5) \times (-6) - 13$

27) Solve. $\frac{6}{14} \times \frac{7}{9} =$

   A. $\frac{1}{3}$

   B. $\frac{5}{26}$

   C. $\frac{21}{63}$

   D. $\frac{1}{7}$

28) A cereal box has a height of 35 centimeters. The area of the base is 120 centimeters. What is the volume of the cereal box?

   Write your answer in the box below.

Common Core Subject Test Mathematics Grade 5

29) Nancy ordered 16 pizzas. Each pizza has 6 slices. How many slices of pizza did Nancy ordered?

A. 46

B. 96

C. 69

D. 116

30) The length of a rectangle is $\frac{6}{25}$ inches and the width of the rectangle is $\frac{5}{18}$ inches. What is the area of that rectangle?

A. $\frac{1}{12}$

B. $\frac{1}{15}$

C. $\frac{30}{400}$

D. $\frac{2}{15}$

31) ABC Corporation earned only $500,000 during the previous year, five–thirds only of the management's predicted income. How much earning did the management predict?

A. $883,000

B. $500,000

C. $300,000

D. $800,000

WWW.MathNotion.Com

Common Core Subject Test Mathematics Grade 5

32) What is the volume of this box?

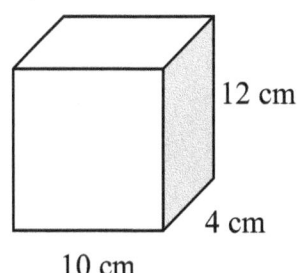

A. 48 cm³

B. 120 cm³

C. 420 cm³

D. 480 cm³

33) How many square feet of tile is needed for 12 feet to 12 feet room?

A. 48 Square Feet

B. 169 Square Feet

C. 121 Square Feet

D. 144 Square Feet

34) Solve. $\frac{1}{3} + \frac{2}{5} - \frac{9}{15} =?$

A. $\frac{7}{15}$

B. $\frac{4}{15}$

C. $\frac{2}{15}$

D. $\frac{2}{5}$

35) Of the 2,300 videos available for rent at a certain video store, 920 are comedies. What percent of the videos are comedies?

A. 60 %

B. 23%

C. 20%

D. 40%

Common Core Subject Test Mathematics Grade 5

36) How many 4 × 4 squares can fit inside a rectangle with a height of 24 and width of 36?

   A. 54

   B. 45

   C. 72

   D. 47

37) William keeps track of the length of each fish that he catches. Following are the lengths in inches of the fish that he caught one day: 80, 41, 14, 27, 18, 10, 61.

   What is the median fish length that William caught that day?

   A. 14 Inches

   B. 41 Inches

   C. 18 Inches

   D. 27 Inches

38) $17 + [21 \times 5] \div 7 = ?$

   Write your answer in the box below.

39) What is the median of these numbers? 27, 38, 28, 29, 47, 36, 23

   A. 47

   B. 29

   C. 28

   D. 36

40) The area of the base of the following cylinder is 45 square inches and its height is 6 inches. What is the volume of the cylinder?

Write your answer in the box below.

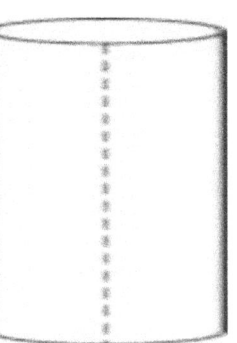

Common Core Practice Test 2

Mathematics

GRADE 5

Released *Month Year*

Common Core Subject Test Mathematics Grade 5

1) Jack added 29 to the product of 24 and 19. What is this sum?

   A. 680

   B. 459

   C. 485

   D. 2,653

2) Joe makes $3.5 per hour at his work. If he works 6 hours, how much money will he earn?

   A. $31

   B. $19

   C. $17

   D. $21

3) Which of the following is an obtuse angle?

   A. 18°

   B. 58°

   C. 148°

   D. 188°

4) What is the value of $5 - 3\frac{1}{8}$?

   A. $\frac{13}{8}$

   B. $1\frac{7}{8}$

   C. $-2\frac{1}{8}$

   D. $2\frac{7}{8}$

Common Core Subject Test Mathematics Grade 5

5) The bride and groom invited 840 guests for their wedding. 630 guests arrived. What percent of the guest list was not present?

A. 45%

B. 15%

C. 45%

D. 25%

6) Frank wants to compare these two measurements.

33.012 kg ☐ 33,012 g

Which symbol should he use?

A. >

B. ≠

C. <

D. =

7) Aria was hired to teach four identical 4th grade math courses, which entailed being present in the classroom 32 hours altogether. At $35 per class hour, how much did Aria earn for teaching one course?

A. $64

B. $280

C. $168

D. $1,240

Common Core Subject Test Mathematics Grade 5

8) In a classroom of 75 students, 45 are male. What percentage of the class is female?

   A. 45%

   B. 25%

   C. 40%

   D. 60%

9) In a party, 4 soft drinks are required for every 18 guests. If there are 270 guests, how many soft drinks are required?

   A. 70

   B. 90

   C. 60

   D. 120

10) You are asked to chart the temperature during an 8–hour period to give the average. These are your results:

    | 7 am: 2 degrees | 11 am: 22 degrees |
    | 8 am: 4 degrees | 12 pm: 30 degrees |
    | 9 am: 20 degrees | 1 pm: 31 degrees |
    | 10 am: 21 degrees | 2 pm: 30 degrees |

    What is the average temperature?

    A. 20

    B. 18

    C. 24

    D. 26

Common Core Subject Test Mathematics Grade 5

11) While at work, Emma checks her email once every 20 minutes. In 12 hours, how many times does she check her email?

   A. 26 Times

   B. 36 Times

   C. 28 Times

   D. 38 Times

12) In a classroom of 80 students, 52 are male. About what percentage of the class is female?

   A. 25%

   B. 45%

   C. 35%

   D. 55%

13) A florist has 987 flowers. How many full bouquets of 21 flowers can he make?

   A. 36

   B. 23

   C. 58

   D. 47

14) What is 3,994.8255 rounded to the nearest tenth?

   A. 3,994.83

   B. 3,994.8

   C. 3,994

   D. 3,994.9

Common Core Subject Test Mathematics Grade 5

15) If a rectangular swimming pool has a perimeter of 120 feet and it is 19 feet wide, what is its area?

   A. 824 square feet

   B. 249 square feet

   C. 668 square feet

   D. 779 square feet

16) What is the volume of the following rectangle prism?

   A. 48 ft^3

   B. 96 ft^3

   C. 184 ft^3

   D. 384 ft^3

   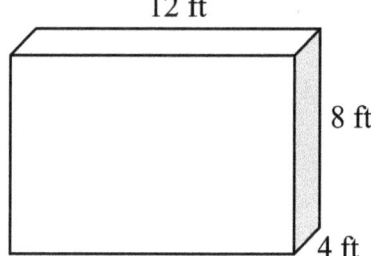

17) A circle has a diameter of 22 inches. What is its approximate circumference? ($\pi = 3.14$)

   A. 56.09 inches

   B. 69.08 inches

   C. 25.65 inches

   D. 58.65 inches

18) If a rectangle is 17 feet by 28 feet, what is its area?

   A. 476

   B. 766

   C. 824

   D. 348

19) How long is the line segment shown on the number line below?

A. 12

B. 14

C. 4

D. 10

20) If a vehicle is driven 41 miles on Monday, 25 miles on Tuesday, and 39 miles on Wednesday, what is the average number of miles driven each day?

A. 35 Miles

B. 38 Miles

C. 33 Miles

D. 32 Miles

21) Peter traveled 270 miles in 9 hours and Jason traveled 400 miles in 8 hours. What is the ratio of the average speed of Peter to average speed of Jason?

A. 3: 5

B. 5: 8

C. 8: 7

D. 7: 5

Common Core Subject Test Mathematics Grade 5

22) If $x = -9$, which equation is true?

   A. $3x(x + 7) = 48$

   B. $7(3 - x) = 84$

   C. $(-3)(2x + 3) = 42$

   D. $6x - 9 = -52$

23) A woman owns a dog walking business. If 3 workers can walk 12 dogs, how many dogs can 5 workers walk?

   A. 15

   B. 20

   C. 25

   D. 30

24) What are the coordinates of the intersection of x-axis and the y-axis on a coordinate plane?

   A. $(3, -3)$

   B. $(-3, 0)$

   C. $(0, 0)$

   D. $(0, -3)$

25) In a triangle ABC the measure of angle ACB is 29° and the measure of angle CAB is 83°. What is the measure of angle ABC?

Write your answer in the box below.

Common Core Subject Test Mathematics Grade 5

26) Which list shows the fractions listed in order from least to greatest?

$$\frac{1}{15}, \frac{1}{5}, \frac{1}{12}, \frac{1}{19}$$

   A. $\frac{1}{12}, \frac{1}{5}, \frac{1}{19}, \frac{1}{15}$

   B. $\frac{1}{15}, \frac{1}{19}, \frac{1}{5}, \frac{1}{12}$

   C. $\frac{1}{5}, \frac{1}{12}, \frac{1}{15}, \frac{1}{19}$

   D. $\frac{1}{19}, \frac{1}{15}, \frac{1}{12}, \frac{1}{5}$

27) A car uses 20 gallons of gas to travel 660 miles. How many miles per gallon does the car get?

   A. 42 miles per gallon

   B. 44 miles per gallon

   C. 33 miles per gallon

   D. 32 miles per gallon

28) Eight out of 64 students had to go to summer school. What is the ratio of students who did not have to go to summer school expressed, in its lowest terms?

   A. $\frac{7}{8}$

   B. $\frac{1}{8}$

   C. $\frac{5}{8}$

   D. $\frac{8}{7}$

Common Core Subject Test Mathematics Grade 5

29) David's motorcycle stalled at the beach and he called the towing company. They charged him $ 5.10 per mile for the first 30 miles and then $5.50 per mile for each mile over 30. David was 42 miles from the motorcycle repair shop. How much was David's towing bill?

   A. $199

   B. $292

   C. $219

   D. $192

30) If one acre of forest contains 180 pine trees, how many pine trees are contained in 38 acres?

   A. 6,840

   B. 840

   C. 84,600

   D. 6,880

31) If 7 garbage trucks can collect the trash of 35 homes in a day. How many trucks are needed to collect in 55 houses?

   A. 9

   B. 8

   C. 12

   D. 11

Common Core Subject Test Mathematics Grade 5

32) A steak dinner at a restaurant costs $7.6. If a man buys a steak dinner for himself and 4 friends, what will the total cost be?

   A. $28.40

   B. $38.00

   C. $42.40

   D. $36.00

33) If Ella needed to buy 4 bottles of soda for a party in which 14 people attended, how many bottles of soda will she need to buy for a party in which 28 people are attending?

   A. 7

   B. 10

   C. 8

   D. 6

34) Julie gives 9 pieces of candy to each of her friends. If Julie gives all her candy away, which amount of candy could have been the amount she distributed?

   A. 389

   B. 556

   C. 378

   D. 429

Common Core Subject Test Mathematics Grade 5

35) Four co–workers contributed $10.58, $14.56, $11.42, and $13.55 respectively to purchase a retirement gift for their boss. What is the maximum amount they can spend on a gift?

   A. $48.22

   B. $52.22

   C. $50.11

   D. $49.11

36) Number 0.078 is equal to what percent?

   A. 0.078%

   B. 0.78%

   C. 7.80%

   D. 78%

37) Which of the following fractions is the largest?

   A. $\frac{9}{14}$

   B. $\frac{4}{7}$

   C. $\frac{5}{8}$

   D. $\frac{7}{9}$

Common Core Subject Test Mathematics Grade 5

38) A rectangular plot of land is measured to be 90 feet by 240 feet. What is its total area?

   A. 21,600 square feet

   B. 18,900 square feet

   C. 23,800 square feet

   D. 19,900 square feet

39) A barista averages making 23 cups of coffee per hour. At this rate, how many hours will it take until she's made 1,472 cups of coffee?

   A. 59

   B. 64

   C. 68

   D. 71

40) Ava needs $\frac{1}{4}$ of an ounce of salt to make 1 cup of dip for fries. How many cups of dip will she be able to make if she has 70 ounces of salt?

   A. 35

   B. 140

   C. 740

   D. 280

"This is the end of Practice Test 2."

Common Core Subject Test Mathematics Grade 5

Chapter 14 : Answers and Explanations

Common Core Practice Tests Answer Key

❋ Now, it's time to review your results to see where you went wrong and what areas you need to improve!

Practice Test - 1						Practice Test - 2					
1	C	16	A	31	C	1	C	16	D	31	D
2	C	17	C	32	D	2	D	17	B	32	B
3	C	18	D	33	D	3	C	18	A	33	C
4	A	19	B	34	C	4	B	19	A	34	C
5	B	20	C	35	D	5	D	20	A	35	C
6	D	21	34	36	A	6	D	21	A	36	C
7	C	22	C	37	D	7	B	22	B	37	D
8	A	23	D	38	32	8	C	23	B	38	A
9	D	24	B	39	B	9	C	24	C	39	B
10	B	25	C	40	270	10	A	25	68	40	D
11	D	26	A			11	B	26	D		
12	D	27	A			12	C	27	C		
13	108	28	4,200			13	D	28	A		
14	A	29	B			14	B	29	C		
15	B	30	B			15	D	30	A		

Common Core Subject Test Mathematics Grade 5

Practice Test 1
Common Core - Mathematics
Answers and Explanations

1) Answer: C.

19 yards long rope is cut into 4 equal parts. Therefore, 19 should be divided by 4. Only option C is NOT 19 divided by 4. (It is 4 divided by 19)

2) Answer: C.

Use area of trapezoid formula.

Area of trapezoid $= \frac{1}{2} \times heigth \times (base\ 1 + base\ 2) \Rightarrow \frac{1}{2} \times 8 \times (5 + 11) = 64$

3) Answer: C.

To find the answer, subtract 45,386 from 45,613.

$45,613 - 45,386 = 227$ miles

4) Answer: A.

Let x be the new price after discount.

$x = 60 \times (100 - 35)\% = 60 \times 65\% = 60 \times 0.65 = 39$

$x = \$39$

5) Answer: B.

7 eggs for 1 cake. Therefore, 910 eggs can be used for $(910 \div 7)$ 130 cakes.

6) Answer: D.

An obtuse angle is any angle larger than 90 degrees. From the options provided, only D (115 degrees) is larger than 90.

7) Answer: C.

Compare the fractions.

$\frac{3}{7} > \frac{1}{4}$

And $\frac{7}{8} > \frac{4}{7}$

$\frac{7}{8} > \frac{3}{7}$; Therefore, $\frac{7}{8}$ is the biggest fraction.

Common Core Subject Test Mathematics Grade 5

8) Answer: A.

Use area of rectangle formula.

area of a rectangle = $width \times length \Rightarrow D = w \times l \Rightarrow w = \frac{D}{l} = \frac{D}{27}$

9) Answer: D.

To list the fractions from least to greatest, you can convert the fractions to decimal.

$\frac{2}{9} = 0.222; \frac{1}{8} = 0.125; \frac{7}{12} = 0.58; \frac{8}{9} = 0.889; \frac{5}{7} = 0.714$

$\frac{1}{8} = 0.125, \frac{2}{9} = 0.222, \frac{7}{12} = 0.58, \frac{5}{7} = 0.714, \frac{8}{9} = 0.889$

Option D shows the fractions in order from least to greatest.

10) Answer: B.

6 multiplied by $\frac{8}{13} = \frac{48}{13} = 3.69$, therefore, only choice B is correct.

11) Answer: D.

Use volume of rectangle formula.

Volume of a rectangle = $width \times length \times heigth \Rightarrow V = 6 \times 8 \times 12 \Rightarrow V = 576$

12) Answer: D.

To find the selling price, multiply the price by (100% − rate of discount).

Then: (360) (100% − 25%) = (360) (0.75) = 270

13) Answer: 108.

First, find the missing side of the trapezoid. The perimeter of the trapezoid below is 44.

Therefore, the missing side of the trapezoid (its height) is:

44 − 7 − 11 − 14 = 44 − 32 = 12

Area of a trapezoid: A = $\frac{1}{2}$ h (b1 + b2)

= $\frac{1}{2}$ (12) (7 + 11) = 108

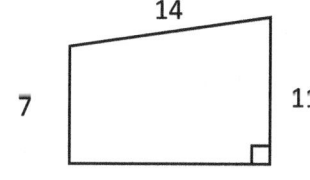

14) Answer: A.

There are 105 cards in the bag and 15 of them are white. Then, 15 out of 105 cards are white. You can write this as: $\frac{15}{105}$. To simplify this fraction, divide both numerator and denominator by 15. Then: $\frac{15}{105} = \frac{1}{7}$

WWW.MathNotion.Com

Common Core Subject Test Mathematics Grade 5

15) Answer: B.

$\frac{9}{40}$ means 9 is divided by 40. The fraction line simply means division or ÷. Therefore, we can write $\frac{9}{40}$ as $9 \div 40$.

16) Answer: A.

Plug in 50 for A in the equations. Only option A works.

$A + 50 = 120$

$70 + 50 = 120$

17) Answer: C.

Use percent formula: part $= \frac{\text{percent}}{100} \times$ whole

$26 = \frac{\text{percent}}{100} \times 40 \Rightarrow 26 = \frac{\text{percent} \times 40}{100} \Rightarrow 26 = \frac{\text{percent} \times 4}{10}$, multiply both sides by 10.

$260 =$ percent $\times 4$, divide both sides by 4. $65 =$ percent

18) Answer: D.

50 miles: 1 hour

425 miles: $425 \div 50 = 8.5$; 8 hours; $0.5 \times 60 = 30$ min

19) Answer: B.

Use area and circumference of circle formula.

Area of a circle $= \pi r^2 \Rightarrow 25\pi = \pi r^2 \Rightarrow r = 5$

Circumference of a circle $= 2\pi r \Rightarrow C = 2 \times 5 \times \pi \Rightarrow C = 10\pi$

20) Answer: C.

1 meter of the rope = 420 grams

50 meters of the rope = $50 \times 420 = 21{,}000$ grams = 21 kilograms

21) Answer is 34.

To solve this problem, divide $4\frac{1}{4}$ by $\frac{1}{8}$.

$4\frac{1}{4} \div \frac{1}{8} = \frac{17}{4} \div \frac{1}{8} = \frac{17}{4} \times \frac{8}{1} = 34$

22) Answer: C.

The question is that number 86,541.654 is how many times of number 8.6541654. The answer is 10,000.

Common Core Subject Test Mathematics Grade 5

23) Answer: D.

Lily eats 8 pancakes in 1 minute ⇒ Lily eats 8 × 6 pancakes in 6 minutes.

Ella eats $3\frac{1}{6}$ pancakes in 1 minute ⇒ Ella eats $3\frac{1}{6}$ × 6 pancakes in 6 minutes.

In total Lily and Ella eat 48 + 19 pancakes in 6 minutes.

24) Answer: B.

Alice drives 78 miles in one hour. Therefore, she drives 3,198 miles in about (3,198 ÷ 78) 41 hours.

25) Answer: C.

10 yards = 10 × 36 = 360 inches

8 feet = 8 × 12 = 96 inches

10 yards 8 feet and 18 inches = 360 inches + 96 inches + 18 inches = 474 inches

26) Answer: A.

Simplify each option provided using order of operations rules.

A. 21 − (− 6) + (− 36) = 21 + 6 − 36 = −9

B. 4 + (− 3) × (− 7) = 4 + 21 = 25

C. − 8 × (− 2) − (− 9) × (− 5) = 16 − 45 = −29

D. (− 5) × (− 6) − 13 = 30 − 13 = 17

Only option A is −9

27) Answer: A.

$\frac{6}{14} \times \frac{7}{9} = \frac{6 \times 7}{14 \times 9} = \frac{42}{126} = \frac{1}{3}$

28) Answer: 4,200.

Use volume of cube formula.

$Volume = base \times height \Rightarrow V = 120 \times 35 \Rightarrow V = 4,200$

29) Answer: B.

1 pizza has 6 slices. 16 pizzas contain (16 × 6) 96 slices.

30) Answer: B.

Use area of rectangle formula.

Area= $length \times width \Rightarrow A = \frac{6}{25} \times \frac{5}{18} \Rightarrow A = \frac{30}{450} = \frac{1}{15}$ inches

Common Core Subject Test Mathematics Grade 5

31) Answer: C.

ABC Corporation's income = $\frac{5}{3}$ management's predicted income.

$500,000 = \frac{5}{3}$ management's predicted income

management's predicted income = $500,000 \times \frac{3}{5}$ = $300,000

32) Answer: D.

Use volume of cube formula.

Volume = $length \times width \times height \Rightarrow V = 10 \times 4 \times 12 \Rightarrow V = 480 \ cm^3$

33) Answer: D.

Find the area of the room which is a square. Use area of square formula.

$S = a^2 \Rightarrow S = 12 \ feet \times 12 \ feet = 144 \ square \ feet$

34) Answer: C.

$\frac{1}{3} + \frac{2}{5} - \frac{9}{15} = \frac{(5 \times 1)+(2 \times 3)-(9 \times 1)}{15} = \frac{2}{15}$

35) Answer: D.

Use percent formula:

$part = \frac{percent}{100} \times whole$

$920 = \frac{percent}{100} \times 2,300 \Rightarrow 920 = percent \times 23 \Rightarrow percent = 40$

36) Answer: A.

Use area of rectangle formula.

$A = a \times b \Rightarrow A = 24 \times 36 \Rightarrow A = 864$

Divide the area by 16 (4 × 4 = 16 squares) to find the number of squares needed.

864 ÷ 16 = 54

37) Answer: D.

Write the numbers in order: 10, 14, 18, 27, 41, 61, 80

Median is the number in the middle. Therefore, the median is 27.

38) Answer: 32.

Use PEMDAS (order of operation):

17 + [21 × 5] ÷ 7 = 17 + (105) ÷ 7 = 17 + (105 ÷ 7) = 17 + 15 = 32

Common Core Subject Test Mathematics Grade 5

39) Answer: B.

Write the numbers in order:

23, 27, 28, 29, 36, 38, 47

Median is the number in the middle. Therefore, the median is 29.

40) Answer: 270.

Use volume of cylinder formula.

Volume = $base \times heigth \Rightarrow V = 45 \times 6 \Rightarrow V = 270$

Common Core Subject Test Mathematics Grade 5

Practice Test 2

Common Core - Mathematics

Answers and Explanations

1) Answer: C.

$29 + (24 \times 19) = 29 + 456 = 485$

2) Answer: D.

1 hour: $3.5

6 hours: $6 \times \$3.5 = \$21$

3) Answer: C.

An obtuse angle is an angle of greater than 90° and less than 180°. From the options provided, only option C (148 degrees) is an obtuse angle.

4) Answer: B.

$5 - 3\frac{1}{8} = \frac{40}{8} - \frac{25}{8} = \frac{15}{8} = 1\frac{7}{8}$

5) Answer: D.

The number of guests that are not present are: $(840 - 630)$ 210 out of $840 = \frac{210}{840}$

Change the fraction to percent: $\frac{210}{840} \times 100\% = 25\%$

6) Answer: D.

Each kilogram is 1,000 grams.

33,012 grams $= \frac{33,012}{1,000} = 33.012$ kilograms. Therefore, two amounts provided are equal.

7) Answer: B.

Aria teaches 32 hours for four identical courses. Therefore, she teaches 8 hours for each course. Aria earns $35 per hour. Therefore, she earned $280 ($8 \times 35$) for each course.

8) Answer: C.

The number of female students in the class is: $(75 - 45)$ 30 out of $75 = \frac{30}{75}$

Change the fraction to percent: $\frac{30}{75} \times 100\% = 40\%$

Common Core Subject Test Mathematics Grade 5

9) Answer: C.

Write a proportion and solve. $\frac{4 \text{ soft drinks}}{18 \text{ guests}} = \frac{x}{270 \text{ guests}} \Rightarrow x = \frac{270 \times 4}{18} \Rightarrow x = 60$

10) Answer: A.

average (mean) $= \frac{\text{sum of terms}}{\text{number of terms}} \Rightarrow$ average $= \frac{2+4+20+21+22+30+31+30}{8} \Rightarrow$ average $= 20$

11) Answer: B.

Every 20 minutes Emma checks her email.

In 12 hours (720 minutes), Emma checks her email (720 ÷ 20) 36 times.

12) Answer: C.

There are 80 students in the class. 52 of the are male and 28 of them are female.

28 out of 80 are female. Then:

$\frac{28}{80} = \frac{x}{100} \rightarrow 2{,}800 = 80x \rightarrow x = 2{,}800 \div 80 = 35\%$

13) Answer: D.

Divide the number flowers by 21: 987 ÷ 21 = 47

14) Answer: B.

Rounding decimals is similar to rounding other numbers. If the hundredths and thousandths places of a decimal is eighty-two or less, they are dropped, and the tenths place does not change. For example, rounding 0.534 to the nearest tenth would give 0.5. Therefore, 3,994.8255 rounded to the nearest tenth is 3,994.8.

15) Answer: D.

Perimeter of rectangle formula:

$P = 2\,(length + width) \Rightarrow 120 = 2\,(l + 19) \Rightarrow l = 41$

Area of rectangle formula: $A = length \times width \Rightarrow A = 41 \times 19 \Rightarrow A = 779$

16) Answer: D.

Use volume of rectangle prism formula.

$V = length \times width \times height \Rightarrow V = 12 \times 4 \times 8 \Rightarrow V = 384$

17) Answer: B.

The diameter of the circle is 22 inches. Therefore, the radius of the circle is 11 inches.

Use circumference of circle formula. $C = 2\pi r \Rightarrow C = 2 \times 3.14 \times 11 \Rightarrow C = 69.08$

Common Core Subject Test Mathematics Grade 5

18) Answer: A.

Use area of rectangle formula.

Area = length × width ⇒ A = 17 × 28 ⇒ A = 476

19) Answer: A.

The line segment is from 4 to −8. Therefore, the line is 12 units.

4 − (−8) = 4 + 8 = 12

20) Answer: A.

average (mean) = $\frac{\text{sum of terms}}{\text{number of terms}}$ ⇒ average = $\frac{41+25+39}{3}$ ⇒ average = 35

21) Answer: A.

Peter's speed = $\frac{270}{9}$ = 30

Jason's speed = $\frac{400}{8}$ = 50

$\frac{\text{The average speed of peter}}{\text{The average speed of Jason}} = \frac{30}{50}$ equals to: $\frac{3}{5}$ or 3: 5

22) Answer: B.

Plug in $x = -9$ in each equation.

A. $3x(x+7) = 48 \rightarrow 3(-9)(-9+7) = (-27) \times (-9+7) = 54$

B. $7(3-x) = 84 \rightarrow 7(3-(-9)) = 7(12) = 84$

C. $(-3)(2x+3) = 42 \rightarrow (-3)(2(-9)+3) = (-3)(-18+3) = 45$

D. $6x - 9 = -52 \rightarrow 6(-9) - 9 = -54 - 9 = -63$

Only option B.

23) Answer: B.

3 workers can walk 12 dogs ⇒ 1 workers can walk 4 dogs.

5 workers can walk (5 × 4) 20 dogs.

24) Answer: C.

The horizontal axis in the coordinate plane is called the $x - axis$. The vertical axis is called the $y - axis$. The point at which the two axes intersect is called the origin. The origin is at 0 on the $x - axis$ and 0 on the $y - axis$.

Common Core Subject Test Mathematics Grade 5

25) Answer: 68°.

All angles in every triangle add up to 180°. Let x be the angle ABC.

Then: $180° = 29° + 83° + x \Rightarrow x = 68°$

26) Answer: D.

In fractions, when denominators increase, the value of fractions decrease and as much as numerators increase, the value of fractions increase. Therefore, the least one of this list is: $\frac{1}{19}$ and the greatest one of this list is: $\frac{1}{5}$

27) Answer: C.

Write a proportion and solve.

20 gallons: 660 miles $\Rightarrow$ 1 gallon: $660 \div 20 = 33$ miles

28) Answer: A.

The students that had to go to summer school is 8 out of $64 = \frac{8}{64} = \frac{1}{8}$

Therefore $\frac{7}{8}$ students did not have to go to summer school.

29) Answer: C.

$5.10 per mile for the first 30 miles. Therefore, the cost for the first 30 miles is:

$30 \times \$5.10 = \$153$

$5.50 per mile for each mile over 30, therefore, 12 miles over 30 miles cost:

$12 \times \$5.50 = \$66$

In total, David pays: $\$153 + \$66 = \$219$

30) Answer: A.

1 acre: 180 pine trees; 38 acres: $180 \times 38 = 6,840$ pine trees

31) Answer: D.

7 garbage trucks can collect the trash of 35 homes. Then, one garbage truck can collect the trash of 5 homes.

To collect trash of 55 houses, 11 ($55 \div 5$) garbage trucks are required.

32) Answer: B.

5 steak dinners = $5 \times \$7.6 = \$38$

WWW.MathNotion.Com

Common Core Subject Test Mathematics Grade 5

33) Answer: C.

Write a proportion and solve.

$\frac{4 \text{ soft drinks}}{14 \text{ guests}} = \frac{x}{28 \text{ guests}} \Rightarrow x = \frac{4 \times 28}{14} \Rightarrow x = 8$

34) Answer: C.

The number of candies should be a whole number! Check each option provided.

A. $389 \div 9 = 43.22$

B. $556 \div 9 = 61.78$

C. $378 \div 9 = 42$

D. $429 \div 9 = 47.67$

Only option C gives a whole number.

35) Answer: C.

They contributed: $10.58 + $14.56 + $11.42 + $13.55 = $50.11 in total, so the maximum amount that they can spend is the sum of their contribution.

36) Answer: C.

$0.078 = \frac{78}{1,000} = \frac{7.8}{100} = 7.8\%$

37) Answer: D.

$\frac{9}{14} = 0.642$; $\frac{4}{7} = 0.571$; $\frac{5}{8} = 0.625$; $\frac{7}{9} = 0.778$

$\frac{7}{9}$ is the largest.

38) Answer: A.

Use area of rectangle formula.

$Area = length \times width \Rightarrow A = 90 \times 240 \Rightarrow A = 21,600$

39) Answer: B.

23 cups: 1 hour. 1,472 cups: $1,472 \div 23 = 64$ hours

40) Answer: D.

Write a proportion and solve. $\frac{\frac{1}{4}}{70} = \frac{1}{x} \Rightarrow x = 70 \times 4 = 280$

"End"

www.ingramcontent.com/pod-product-compliance
Lightning Source LLC
Chambersburg PA
CBHW080439110426
42743CB00016B/3216